The Small Business Innovation Research Program: CHALLENGES AND OPPORTUNITIES

CHARLES W. WESSNER, *Editor*

NATIONAL ACADEMY PRESS
Washington, D.C.

NATIONAL ACADEMY PRESS • 2101 Constitution Ave., N.W. • Washington, D.C. 20418

The National Academy of Sciences is a private, nonprofit society of distinguished scholars engaged in scientific and engineering research, dedicated to the furtherance of science and technology and to their use for the general welfare. Upon the authority of the charter granted to it by Congress in 1863, the Academy has a mandate that requires it to advise the federal government on science and technical matters. Dr. Bruce Alberts is president of the National Academy of Sciences.

The National Academy of Engineering was established in 1964, under the charter of the National Academy of Sciences, as a parallel organization of outstanding engineers. It is autonomous in its administration and in the selection of its members, sharing with the National Academy of Sciences the responsibility for advising the federal government. The National Academy of Engineering also sponsors engineering programs aimed at meeting national needs, encourages education and research, and recognizes the superior achievements of engineers. Dr. William A. Wulf is president of the National Academy of Engineering.

The Institute of Medicine was established in 1970 by the National Academy of Sciences to secure the services of eminent members of appropriate professions in the examination of policy matters pertaining to the health of the public. The Institute acts under the responsibility given to the National Academy of Sciences by its congressional charter to be an adviser to the federal government and, upon its own initiative, to identify issues of medical care, research, and education. Dr. Kenneth I. Shine is president of the Institute of Medicine.

The National Research Council was organized by the National Academy of Sciences in 1916 to associate the broad community of science and technology with the Academy's purposes of furthering knowledge and advising the federal government. Functioning in accordance with general policies determined by the Academy, the Council has become the principal operating agency of both the National Academy of Sciences and the National Academy of Engineering in providing services to the government, the public, and the scientific and engineering communities. The Council is administered jointly by both Academies and the Institute of Medicine. Dr. Bruce Alberts and Dr. William A. Wulf are chairman and vice chairman, respectively, of the National Research Council.

Limited copies are available from:
Board on Science, Technology,
and Economic Policy
National Research Council
1055 Thomas Jefferson, N.W.
Washington, D.C. 20007
202-334-2200

Additional copies are available for sale from:
National Academy Press
Box 285
2101 Constitution Ave., N.W.
Washington, D.C. 20055
800-624-6242
202-334-3313 (in the Washington Metropolitan Area)

International Standard Book Number 0-309-06198-9

Printed in the United States of America

For the National Research Council, this project was overseen by the Board on Science, Technology and Economic Policy (STEP), a standing board of the NRC established by the National Academies of Sciences and Engineering and the Institute of Medicine in 1991. The mandate of the STEP Board is to integrate understanding of scientific, technological, and economic elements in the formulation of national policies to promote the economic well-being of the United States. A distinctive characteristic of STEP's approach is its frequent interactions with public and private sector decisionmakers. STEP bridges the disciplines of business management, engineering, economics, and the social sciences to bring diverse expertise to bear on pressing public policy questions. The members of the STEP Board* and the NRC staff are listed below:

*Membership as of February 1998.

Staff

Stephen A. Merrill
Executive Director

Charles W. Wessner
Program Director

John B. Horrigan
Consultant

John Oldfield
Program Associate

Craig M. Schultz
Program Associate

Ryan L. Catteau
Program Associate

Laura T. Holliday
Program Associate

STEERING COMMITTEE FOR GOVERNMENT-INDUSTRY PARTNERSHIPS FOR THE DEVELOPMENT OF NEW TECHNOLOGIES

Gordon Moore, *Chair*
Chairman Emeritus
Intel Corporation

M. Kathy Behrens
Managing Partner
BancBoston Robertson Stephens
and STEP Board

Gordon Binder
Chief Executive Officer
Amgen, Inc.

Michael Borrus
Co-Director
Berkeley Roundtable on International Economics
University of California

Iain Cockburn
Professor of Commerce and Business Administration
University of British Columbia

Kenneth Flamm
Dean Rusk Chair in International Affairs
LBJ School of Public Affairs
University of Texas at Austin

James F. Gibbons
Professor of Engineering
Stanford University
and STEP Board

William J. Spencer, *Vice-Chair*
Chairman, SEMATECH
and STEP Board

W. Clark McFadden
Partner
Dewey Ballantine

Mark B. Myers
Senior Vice President
Xerox Corporation
and STEP Board

Richard Nelson
George Blumnethal Professor of International and Public Affairs
Columbia University

Charles Trimble
Vice Chairman
Trimble Navigation

John P. Walker
Chairman and Chief Executive Officer
Axys Pharmaceuticals, Inc.

Patrick Windham
Adjunct Professor
Stanford University
Consultant
Science and Technology Policy

Project Staff

Charles W. Wessner
Study Director

John B. Horrigan
Consultant

John Oldfield
Program Associate

Craig Schultz
Administrative Assistant

Ryan Catteau
Program Associate

Laura T. Holliday
Program Associate

NATIONAL RESEARCH COUNCIL
BOARD ON SCIENCE, TECHNOLOGY, AND ECONOMIC POLICY

Sponsors

The National Research Council gratefully acknowledges the support of the following sponsors:

U.S. Department of Defense

U.S. Department of Energy

National Cancer Institute

National Science Foundation

National Institute of Standards and Technology

National Aeronautics and Space Administration

Sandia National Laboratories

Electric Power Research Institute

Advanced Micro Devices

Merck and Company

Milliken Industries

Motorola

Nortel

Procter and Gamble

Silicon Valley Group Incorporated

Contents

I

PREFACE

Preface

Driven both by the exigencies of national defense and the requirements of transportation and communication across the American continent, the federal government has played an instrumental role in the development of new production techniques and technologies from the earliest years of the republic. To do so, the government has often turned to individual entrepreneurs with innovative ideas. For example, in 1798, the federal government laid the foundation for the first machine tool industry with a contract to the inventor, Eli Whitney, for interchangeable musket parts.[1] A few decades later, in 1842, a hesitant Congress appropriated funds to demonstrate the feasibility of Samuel Morse's telegraph.[2] Both men fostered significant innovations which led to whole new industries.

Despite Whitney's ultimate success and the enormous consequences of Morse's ground-breaking innovation, the appropriate role of the government in

[1] Whitney missed his first delivery date and encountered substantial cost overruns. However, his invention of interchangeable parts, and the machine tools to make them, was ultimately successful. The muskets were delivered and the foundation of a new industry was in place. As early as the 1850s, the United States had begun to export specialized machine tools to the Enfield Arsenal in Great Britain. The British described the large-scale production of firearms, made with interchangeable parts, as "the American system of manufacturers" David C. Mowery and Nathan Rosenberg, *Paths of Innovation: Technological Change in 20th Century America*. Cambridge University Press, New York, 1998, p 6.

[2] For a discussion of Samuel Morse's 1837 application for a grant and the congressional debate, see Irwin Lebow, *Information Highways and Byways*. Institute of Electrical and Electronics Engineers, New York, 1995, pp. 9-12. For a more detailed account, see Robert Luther Thompson, *Wiring a Continent: The History of the Telegraph Industry in the United States 1823-1836*. Princeton University Press, Princeton, N.J., 1947.

the economy has remained a source of debate and discussion in the United States to this day. Perhaps the earliest articulation of the government's nurturing role with regard to the composition of the economy was Alexander Hamilton's 1791 *Report on Manufacturers* in which he urged an activist approach by the federal government. At the time, Hamilton's views were controversial, although subsequent U.S. policy has largely reflected his beliefs.

During both the nineteenth and twentieth centuries, the federal government has had an enormous impact on the structure and composition of the economy through infrastructure development, regulation, procurement, and a vast array of policies to support industrial and agricultural development.[3] Between World War I and World War II, these policies included support for the development of key industries, which we would now call dual-use, such as radio and aircraft frames and engines. The requirements of World War II generated a huge increase in government procurement and support for high-technology industries. At the industrial level, there were "major collaborative initiatives in pharmaceutical manufacturing, petrochemicals, synthetic rubber, and atomic weapons."[4] An impressive array of weapons based on new technologies was developed during the war, ranging from radar and improved aircraft, to missiles and, not least, the atomic bomb. The government also played a central role in the creation of the first electronic digital computer, the ENIAC.[5] Following the war, the federal government began to fund basic research at universities on a significant scale, first through the Office of Naval Research and later through the National Science Foundation.[6]

During the Cold War, the United States continued to emphasize technological superiority as a means of ensuring U.S. security. Government funds and costplus contracts helped to support systems and enabling technologies such as semiconductors and new materials, radar, jet engines, missiles, and computer hard-

[3] Examples abound. The government played a key role in the development of the U.S. railway network, growth of agriculture through the Morrill Act (1862) and the agricultural extension service, and support of industry through the National Bureau of Standards (1901). See Richard Bingham, *Industrial Policy American Style: From Hamilton to HDTV*. New York: M.E. Sharpe, 1998 for a comprehensive review.

[4] David Mowery, "Collaborative R&D: how effective is it?" *Issues in Science and Technology.* 1998, p. 37.

[5] Kenneth Flamm, *Creating the Computer*. Washington, DC: The Brookings Institution, 1988, chapters 1-3.

[6] The National Science Foundation was initially seen as the agency that would fund basic scientific research at universities after World War II. However, disagreements over the degree of Executive Branch control over the NSF delayed passage of its authorizing legislation until 1950, even though the concept for the agency was first put forth in 1945 in Vannevar Bush's report *Science: The Endless Frontier*. The Office of Naval Research bridged the gap in basic research funding during these years. For an account of the politics of the NSF's creation, see G. Pascal Zachary, *Endless Frontier: Vannevar Bush, Engineer of the American Century*, New York: The Free Press, 1997, pp. 231. See also Daniel Lee Kleinman, *Politics on the Endless Frontier: Postwar Research Policy in the United States*, Durham, NC: Duke University Press, 1995.

ware and software.[7] In the post-Cold War period, the evolution of the American economy continues to be profoundly marked by the interaction of government-funded research and innovative entrepreneurs. Government support in areas such as microelectronics, robotics, biotechnology, the human genome, and the development of ARPANET, the forerunner of today's Internet, are providing the underpinnings of a new economy. Individual entrepreneurs and researchers often played leading roles in developing new approaches and new businesses to exploit these research investments.[8]

Despite the important role the U.S. government has played in the development of the American economy, there is little consensus concerning the principle of government participation and there is often considerable debate about the appropriate mechanisms of participation. At the same time, in light of the rising costs, substantial risks and the breadth of potential applications of new technologies, some believe that a supportive policy framework by the government is necessary if new, welfare-enhancing and wealth-generating technologies are to be developed and brought to the market.

Since 1991 the National Research Council's Board on Science, Technology, and Economic Policy (STEP) has undertaken a program of activities to improve policy makers' understanding of the interconnections of science, technology, and economic policy and their importance for the American economy and its international competitive position. The Board's activities have corresponded with increased policy recognition of the importance of technology to economic growth. The new economic growth theory emphasizes the role of technology creation, which is believed to be characterized by significant growth externalities.[9] A consequence of the renewed appreciation of growth externalities is recognition of the economic geography of economic development. With growth externalities coming about in part from the exchanges of knowledge among innovators, certain regions become centers for particular types of high growth activities. Innovators are able to take advantage of the tacit knowledge available in such centers to address technology and other business development issues.[10]

In addition, some economists have suggested limitations to traditional trade theory, particularly with respect to the reality of imperfect international competi-

---

[7] For an excellent review of the role of government support in nurturing the computer industry, see National Research Council, *Funding a Revolution: Government Support for Computing Research*, National Academies Press, Washington, D.C., 1999.

[8] David B. Audretsch and Roy Thurik, *Innovation, Industry, Evolution, and Employment*. Cambridge University Press, 1999.

[9] Paul Romer, "Endogenous technological change," *Journal of Political Economy*, Vol. 98, 1990, p. 71-102. See also Gene Grossman and Elhanan Helpman, *Innovation and Growth in the Global Economy*, Cambridge, Mass., MIT Press, 1993.

[10] Paul Krugman, *Geography and Trade*, Cambridge, Mass., MIT Press, 1991, p . 23, points out how the British economist Alfred Marshall initially observed in his classic *Principles of Economics* how geographic clusters of specific economic activities arose from the exchange of "tacit" knowledge among business people.

tion.[11] Recent economic analysis suggests that high-technology is often characterized by increasing rather than decreasing returns, justifying to some the proposition that governments can capture permanent advantage in key industries by providing relatively small, but potentially decisive support to bring national industries up the learning curve and down the cost curve. The increasing recognition of the dynamic element of technological innovation, in particular its cumulative nature, has provided an intellectual underpinning for strategic trade concepts that emphasize the dynamic nature of international competition in high-technology industries.[12]

PROJECT ORIGINS

The growth in government programs to support high technology industry within national economies and their impact on international science and technology cooperation and on the multilateral trading system are of considerable interest worldwide. Accordingly, these topics were taken up by STEP in a study carried out in conjunction with the Hamburg Institute for Economic Research and the Institute for World Economics in Kiel which produced the 1996 report, *Conflict and Cooperation in National Competition for High-Technology Industry.* One of the principal recommendations for further work emerging from that study was a call for an analysis of the principles of effective cooperation in technology development, to include lessons from national and international consortia, including eligibility standards and assessments of what new cooperative mechanisms might be developed to meet the challenges of international cooperation in high-technology products.[13]

In many high-technology industries, the burgeoning development costs for new technologies, the dispersal of technological expertise, and the growing importance of regulatory and environmental issues have provided powerful incentives for public-private cooperation. Notwithstanding the unsettled policy environment in Washington, collaborative programs have steadily expanded.[14]

---

[11] Paul Krugman, *Rethinking International Trade*, Cambridge, Mass., MIT Press, 1990.

[12] For a discussion of governments' efforts to capture new technologies and the industries they spawn for their national economies, see National Research Council, *Conflict and Cooperation in National Competition for High-Technology Industry*, National Academy Press, Washington, D.C. 1996, pp. 28-40. For a critique of these efforts, see Paul Krugman, *Peddling Prosperity: Economic Sense and Nonsense in an Age of Diminished Expectations.* New York: W.W. Norton, 1995.

[13] The summary report of the project (National Research Council, *op.cit.*) recommends further analytical work concerning principles for effective cooperation in technology development (see Recommendation 24, p. 8). More recently, David Mowery has noted the rapid expansion of collaborative activities and emphasized the need for comprehensive assessment. David Mowery, "Collaborative R&D: how effective is it?" *op. cit.*, p. 44.

[14] In addition to programs such as SBIR, SEMATECH, and ATP, other legislative initiatives sought to encourage cooperation and improve the payoff from federal R&D. Examples include the Stevenson-Wydler Technology Innovation Act (1980), the Bayh-Dole University and Small Business Patent Act (1980), the National Cooperative Research Act (1984), and the Federal Technology Transfer Act (1986).

During the Reagan administration, the Small Business Innovation Research program (SBIR) was created as a way to use the innovative capacity of small business as a means of using federal R&D dollars more effectively. To meet unprecedented challenges in the semiconductor industry, the SEMATECH consortium was established, although only after much debate.[15] In the Bush administration, Congress first funded the Advanced Technology Program (ATP) in the National Institute of Standards and Technology and the Advanced Battery Consortium was created. The Clinton administration came to office with an emphasis on civilian technology programs, substantially expanding the ATP and creating the Technology Reinvestment Program (TRP), and the Partnership for the Next Generation Vehicle (PNGV).[16] The rapid expansion of these cooperative programs encountered significant opposition, rekindling the national debate on the appropriate role of the government in fostering new technologies. Indeed, broader philosophical questions about the appropriate role for government in collaborating with industry have tended to obscure the need for policy makers to draw lessons from current and previous collaborative efforts.

Given the considerable change in federal research and development budgets since the end of the Cold War, and the reduced role of many centralized laboratories in the private sector, government-industry collaboration is of growing importance, yet it has seen remarkably little objective analysis. At one level, analysis may contribute to a better appreciation of the role of collaboration between government and industry in the development of the U.S. economy. Writing twenty years ago, one well-known American economist observed that Americans are still remarkably uninformed about their long history of policies aimed at stimulating innovation.[17] Today, many Americans appreciate the contribution of tech-

---

[15] For a review of SEMATECH, see the National Research Council, 1996, *op.cit.*, pp. 141-151. For one of the most comprehensive assessments of SEMATECH, see John B. Horrigan, "Cooperating Competitors: A Comparison of MCC and SEMATECH," monograph, National Research Council, Washington, D.C. forthcoming.

[16] For an analysis of ATP, see Christopher T. Hill, "The Advanced Technology Program: opportunities for enhancement," in Lewis Branscomb and James Keller, eds. *Investing in Innovation: Creating a Research and Innovation Policy.* Cambridge, Mass., MIT Press, 1998, pp. 143-173. For an excellent analysis of the TRP, see Jay Stowsky, "Politics and Policy: The Technology Reinvestment Program and the Dilemmas of Dual Use." Mimeo, University of California, 1996. See also, Linda R. Cohen, "Dual-use and the Technology Reinvestment Project." in Branscomb and Keller, *op.cit.*, pp. 174-193. For PNGV, see National Research Council, *Review of the Research Program of the Partnership for a New Generation of Vehicles: Third Report.* Washington, D.C.: The National Academy Press, 1997. See *Effectiveness of the United States Advanced Battery Consortium as a Government-Industry Partnership*, National Academy Press, Washington, D.C. 1998.

[17] Otis L. Graham, *Losing Time: The Industrial Policy Debate.* Harvard University Press, Cambridge, Mass., 1992, p. 250. Graham cites Richard Nelson's observations at the end of the Carter Administration. The situation may not have improved. Writing in 1994, James Fallows makes a similar observation (see *Looking into the Sun: The Rise of the New East Asian Economic and Political System.* New York: Pantheon Books, 1994, p. 196). See also Thomas McCraw's "Mercantilism and the market: antecedents of American industrial policy," in *The Politics of Industrial Policy,* Claude E.

nology to the current period of robust economic growth.[18] Yet there is little evidence that Americans are aware of the key contributions of federal support for technological innovation, from radio to the Internet.

Leaving aside the desirability of having a better understanding of the role of partnerships in fostering new technologies, one compelling argument for assessment is the simple fact that government intervention in the market is fraught with risk. There are cases of major success, such as federal support to the computer or semiconductor industries, where the Department of Defense served as a source of R&D and as a reliable, early buyer of products.[19] There are also cases of major frustration. Landmarks would include projects such as the Supersonic Transport and the Synfuels Corporation.[20] Regular assessment is vital to ensure continued technical viability, though cost-sharing requirements can be an effective safeguard. Assessment can also help avoid "political capture" of projects, especially large commercial demonstration efforts.[21] Even successful collaborations face the challenge of adapting programs to rapidly changing technologies.[22] Assessment thus becomes a means of keeping programs relevant. Assessment can also have the virtue of reminding policymakers of the need for humility before the "black box" of innovation. As one observer notes, "experience argues for hedged commitments, constant reappraisal, maintenance of options, pluralism of advice and decision makers."[23]

From an international perspective, understanding the benefits and challenges

---

Barfield and William A. Schambra, eds., American Enterprise Institute for Public Policy Research, Washington, D.C., 1986, pp. 33-62.

[18] For a review of support for computing and the Internet, see National Research Council, *Funding a Revolution: Government Support for Computing Research*, National Academies Press, Washington, D.C., 1999, *op. cit.* Chapter 7.

[19] *Ibid.* See also Graham, *op.cit.*, p. 2.

[20] See Linda R. Cohen and Roger G. Noll, *The Technology Pork Barrel*, The Brookings Institution, Washington, D.C., 1991, pp. 97, 178, 259-320, 217-258. An interesting review of technology development programs, mainly from the 1970s, the analysis is less negative than the title suggests. Indeed, the volume identifies some successful R&D projects such as the photovoltaic electricity program. The recent analyses by the Academies of government support for the computing also provide a valuable perspective on the importance and success of sustained government support.

[21] Cohen and Noll stress that political capture by distributive congressional politics and industrial interests are one of the principal risks for government-supported commercialization projects. In cases such as the Supersonic Transport project, they extensively document the disconnect between declining technical feasibility and increasing political support (see *op.cit.*, p. VII and pp. 242-257.

[22] One of the strengths of SEMATECH was its ability to redefine goals in the face of changing conditions. See National Research Council, 1996, *Conflict and Cooperation*, p. 148. See also Grindley, et. al., "SEMATECH and collaborative research: lessons in the design of high-technology consortia." *Journal of Policy Analysis and Management,* 1994, p. 724 and Peter Grindley and William Spencer, "SEMATECH after five years: high-tech consortia and U.S. competitiveness," *California Management Review*. Vol. 35, no. 4, pp. 9-33 and Horrigan, "Cooperating Competitors," *op. cit.*

[23] Otis Graham, *op.cit.*, p. 251. Graham is referring to work by Richard R. Nelson in *Government and Technological Progress*, Pergamon Press, New York, 1982, p. 454-455.

of these programs is also important insofar as they have been, and remain, a central element in the national development strategies of both industrial and industrializing countries. Governments have shown a great deal of imagination in their choice of mechanisms designed to support industry. They have adopted a wide range of policies from trade regulations designed to protect domestic products from foreign competition to tax rebates intended to stimulate the export of selected domestic products. They provide government R&D funding for enterprises of particular interest, and sometimes give overt support through direct grants, loans, and equity investments or more opaque support through mechanisms such as tax deferral.[24] Data collected by the Paris-based Organization for Economic Cooperation and Development (OECD) suggest that worldwide government expenditures on support for high-technology industries involve significant resources and are increasingly focused on what policy makers consider to be strategic industries.[25]

The United States is an active, if unavowed, participant in this global competition, at both the state and the federal level. Indeed, the United States has a remarkably wide range of public-private partnerships in high-technology sectors.[26] In addition to the well-known cases mentioned above, there are public-private consortia of many types. They can be classified in a number of ways, such as by the economic objective of the partnership, that is, to leverage the social benefits associated with federal R&D activity, to enhance the position of a national industry, or to deploy industrial R&D to meet military or other government missions.[27] The program taken up in this symposium, the Small Business Innovation Research Program (SBIR), falls under the latter category.

PROJECT STEERING COMMITTEE

The continual importance of government-industry collaboration underscores the need for better understanding of the opportunities and limitations of these programs and the conditions most likely to ensure success. Reflecting the interest of policy makers in this topic, the STEP Board initiated the project on "Government-Industry Partnerships for the Development of New Technologies," which

---

[24] National Research Council, 1996, *op.cit.*, Box B., pp. 39-40.

[25] *Ibid.* Concerning support for small business, the OECD gives a positive review of U.S. programs. See OECD, *Technology, Productivity, and Job Creation: Best Policy Practices.*, Paris: OECD, 1997. P. 21.

[26] See Chris Coburn and Dan Bergland, *Partnerships*. Batelle Press, Columbus, Ohio, 1995.

[27] See Albert Link, "Public/Private Partnerships as a Tool to Support Industrial R&D: Experiences in the United States." Paper prepared for the working group on Innovation and Technology Policy of the OECD Committee for Science and Technology Policy, Paris, 1998, p. 20. Partnerships can also be differentiated by the nature of public support. Some partnerships involve a direct transfer of funds to an industry consortium. Others focus on the shared use of infrastructure, such as laboratory facilities.

has benefited from broad support among federal agencies. These include the U.S. Department of Defense, the U.S. Department of Energy, the National Science Foundation, the National Cancer Institute, the National Aeronautics and Space Administration, the National Institute of Standards and Technology, as well as a diverse group of private corporations. To carry out this analysis, the STEP Board has assembled a distinguished multidisciplinary steering committee for government-industry partnerships, listed in the front of this report. The committee's principal tasks are to provide overall direction and relevant expertise in the assessment of the issues raised by the project. At the conclusion of the project, the steering committee will develop a consensus report outlining their findings and recommendations on the issues reviewed by the project.

As a basis for the consensus report, the steering committee has undertaken to commission research and convene a series of fact-finding meetings in the form of workshops, symposia, and conferences as a means of informing its deliberations. This symposium, and the proceedings reported here, represent one element of this fact finding effort. It was the first in a series of fact-finding meetings convened under the auspices of the STEP Board and under the direction of the steering committee.[28]

A number of distinguished individuals deserve recognition for their willingness to review this report. These individuals were chosen for their diverse perspectives and technical expertise, in accordance with procedures approved by the NRC's Report Review Committee. The purpose of this independent review is to provide candid and critical comments that will assist the institution in making the published report as sound as possible and to ensure that the report meets institutional standards for objectivity, evidence, and responsiveness to the study charge. The review comments and draft manuscript remain confidential to protect the integrity of the deliberative process. We wish to thank the reviewers for their participation in the review process: James Turner, House Science Committee; Robert B. Archibald, The College of William and Mary; David B. Audretsch, Indiana University; Albert N. Link, University of North Carolina at Greensboro; John T. Scott, Dartmouth College; and Peter Cahill, BTRTC. We are especially grateful for the contributions of the Review Coordinator, Gerald Dinneen. Although these individuals have provided constructive comments and suggestions, it must be emphasized that responsibility for the final content of this report rests entirely with the STEP Board and the NRC.

[28] Other volumes in this series include a companion volume on SBIR entitled *The Small Business Innovation Research Program: An Assessment of DoD's Fast Track Initiative,* National Academy Press, Washington, D.C., 1999; *A Review of the Sandia Science and Technology Park Initiative.* National Academy Press, Washington, D.C., 1999; and *The Advanced Technology Program: Challenges and Opportunities,* National Academy Press, 1999. With respect to international cooperation, the series includes *New Vistas in Transatlantic Science and Technology Cooperation,* National Academy Press, Washington, D.C., 1999.

Given the quality and number of presentations at this symposium, summarizing the proceedings was a challenge. Every effort was made to capture the main points made during presentations and ensuing discussions, within the constraints imposed by the nature of a symposium summary. We apologize in advance for inadvertent errors and omissions in the summary. We also take this opportunity to thank our speakers and participants for making their experience and expertise available to the Academies and our project. Finally, we emphasize that the proceedings that follow do not make findings or recommendations; rather, they seek to capture the different perspectives of the participants and observers of the SBIR program.

Charles W. Wessner

II

INTRODUCTION

Introduction

Small business is widely believed to be a significant source of innovation and associated employment growth in the American economy, a perception that has considerable basis in fact.[1] Certainly in the 19th century, the individual inventor played a central role in American economic development. More recently, the role of the small start-up firms in regions such as Silicon Valley have reinforced the notion that small business is an important driver of economic growth.

Yet the question of firm size and economic growth has been subject of debate for much of this century. The early part of the century was marked by the rise of the large-scale enterprise in the United States, and the conventional wisdom held that large firms had compelling advantages in most performance measures, from profitability to productivity. It was widely accepted that large firms could operate at sufficient levels of scale to produce efficiently and generate the resources to develop new innovations that would perpetuate market dominance. In 1950 Schumpeter, while pointing to the small entrepreneur as the vanguard of the wave of "creative destruction" that spurred innovation, nonetheless posited that large firms, with substantial resources available for R&D, would come to dominate

[1] A recent report by the Organization for Economic Cooperation and Development (OECD) notes that small and medium-sized enterprises are attracting the attention of policymakers, not least because they are seen as major sources of economic vitality, flexibility, and employment. Small business is especially important as a source of new employment, accounting for a disproportionate share of job creation. See OECD, *Small Business Job Creation and Growth: Facts, Obstacles, and Best Practices*. Paris, 1997. For specifics on job growth, see Steven J. Davis, John Haltiwanger, and Scott Schuh, "Small Business and Job Creation: Dissecting the Myth and Reassessing the Facts," *Business Economics*. Vol. 29, no. 3, 1994, pp. 113-22.

capitalist economies.[2] Galbraith later argued that the source of innovation was more plausibly the large firm, which has the resources available to invest at sufficient scale, not the individual innovator.[3]

Concentration and centralization in research and development also characterized the beginning years of the 20th century, and seemed consistent with the ideas about firm size and innovation hypothesized by Schumpeter and Galbraith. The great corporate research laboratories were established at companies such as DuPont, General Electric, and AT&T. In the post-war years, RCA's Sarnoff Laboratory was established, and IBM's Yorktown lab and Bell Laboratories enjoyed their heyday, generating innovations in computing and communications that have had profound effects on the U.S. economy and lifestyle.

By the 1970s, most data indicate that the story began to change, with small-firm growth accelerating. From 1975 to 1984, employment in firms with between 20 and 99 workers grew by 3.64 percent annually, while employment at firms with more than 1,000 workers grew at only one-third that rate, or 1.25 percent.[4] From 1980 to 1987, the average real GNP per firm decreased by 14 percent, from $245,000 to $200,000.[5] As *The Economist* noted in 1989, large firms are shrinking in size and small ones are proliferating; in terms of the source of employment growth "[T]he trend of a century is being reversed."[6] With respect to large research laboratories, as Rosenbloom and Spencer have noted, a similar reduction in size has occurred, as IBM's Yorktown facility was severely downsized in the 1990s and as the breakup of the Bell System in the 1980s changed the character of Bell Labs.[7] Investment in long-term R&D is seen by many as the primary casualty of these changes.[8]

Even before the break up of the large R&D laboratories, there was a growing recognition of the role of small business in furthering technological innovation. The 1980s saw the emergence of rapidly growing companies such as Microsoft

---

[2] Joseph Schumpeter, *Capitalism, Socialism and Democracy*. New York: Harper and Row, 1950, p. 110.

[3] John Kenneth Galbraith, *The New Industrial State*. Boston: Houghton Mifflin, 1957. In fact, evidence from the post-war era seemed to support these notions; from 1947 to 1980, the average real gross national product per firm grew from $150,000 to $245,000; see Zoltan J. Acs and David B. Audretsch, *Innovation and Small Business*. Cambridge, Mass: MIT Press, 1991, p 4.

[4] *Ibid.*

[5] *Ibid.*, p. 3.

[6] "The Rise and Fall of America's Small Firms," *The Economist,* January 21, 1989, pp. 73-74.

[7] Richard Rosenbloom and William Spencer. *Engines of Innovation: U.S. Industrial Research at the End of an Era*. Boston: Harvard Business Press, 1996. Irwin Lebow supports this view, observing that in the opinion of many, the most significant change brought about by the AT&T divestiture was that Bell Laboratories no longer operates under conditions as favorable to the pursuit of fundamental research, the results of which will not be evident for some time in the future. *Information Highways and Byways, op.cit.* pg. 157.

[8] *Ibid.*

and Apple Computing. That decade also saw the rapid growth of the U.S. venture capital industry, which facilitated the contribution of small firms in exploiting the commercial potential of promising new technologies.[9] To some extent, science and technology policy in the 1980s and 1990s has reflected this emphasis on the innovative role of small business.[10]

In addition to the growing recognition of the importance of small business for innovation and employment, there is also today a better understanding of the problems that small businesses face in financing growth. One type of problem has to do with imperfect information in the market for start-up financing. As Lerner notes, asymmetries in information between entrepreneurs and financiers are likely to work to the disadvantage of small firms.[11] Even though providers of funds have strong incentives to gather information about the small business in which they may be investing, the entrepreneur—especially in technology startups—is likely to be the only person with in-depth knowledge of the technology and the market opportunity. And that knowledge is likely to be insufficient to perfectly predict potential payoffs. The result is "statistical discrimination"—it makes sense for financiers to withhold funds even for promising opportunities because it is too costly, and often impossible, to gather the information to assess potential payoffs.[12]

A second problem involves the appropriability of R&D results. The economics literature has long recognized that knowledge is "leaky", that is, new knowledge often transcends the boundaries of firms and intellectual property protection, so that the creator of that knowledge cannot fully capture the economic value of the knowledge through the price system.[13] Moreover, several case-study analyses of small businesses suggest that appropriability problems may be

---

[9] For discussion of the relationship between innovative activity and firm size see Acs and Audretsch, *op.cit.* chap. 3. They maintain that it is important to recognize that small firms are not necessarily more innovative than large firms. The relative contribution depends on the sector, market structure, capital intensity, and rate of innovation. Acs and Audretsch emphasize that both large and small firms bring advantages to the innovative process. Large firms have the resources for long-term R&D investments and benefit from substantial advantages, such as economies of scale, investment, and the market power necessary to recoup R&D investments. Small firms tend to have a higher tolerance for risk, are characterized by rapid decision-making, and often focus on innovative activity as a core strategy. *op.cit.* pg. 39-41.

[10] David Audretsch and Roy Thurik, *Innovation, Industry Evolution, and Employment.* Cambridge: Cambridge University Press, 1999.

[11] See Joshua Lerner, "Public Venture Capital: Rationale and Evaluation" in the Annex of this volume.

[12] *Ibid.*

[13] See Edwin Mansfield, "How Fast Does New Industrial Technology Leak Out?" *Journal of Industrial Economics.* Vol. 34, no. 2, pp. 217-224 for a discussion of the rationale for firms locating R&D facilities in foreign countries, one of which is to absorb leaky R&D information from firms in those countries.

particularly acute for small businesses.[14] In other words, R&D-generated innovations may escape the organizational walls of small firms with relatively greater ease than large businesses.[15] At the same time, ideas not valued and pursued in one firm are often the reason an entrepreneur starts a new firm.[16]

ORIGINS OF THE SMALL BUSINESS INNOVATION RESEARCH PROGRAM

By the late 1970s, the growing prominence of small businesses—and the recognition that they were playing an increasing role in innovation—led to policy responses by the Federal government. As Roland Tibbetts, generally regarded as one of SBIR's founding fathers, recalls in panel 1, federal commissions dating to the 1960s recommended directing R&D funds to small businesses, but such recommendations were not acted upon due to opposition from other recipients of federal R&D funding. It was not until 1976 that any formal action was taken to channel federal R&D funds to small business. That year the National Science Foundation increased the share of its funds going to small business. Small firms were enthused with this initiative, and proceeded to lobby other agencies to follow NSF's lead.

There was no immediate response to these efforts and small businesses therefore took their case to Congress and higher levels of the Executive branch. One result was a White House Conference on Small Business held in January, 1980 that explored specific ways to respond to small business' concerns. From that Conference came a recommendation for legislation that eventually became the Small Business Innovation Development Act of 1982, the bill that authorized the SBIR program. One reason the conference's recommendations received a sympathetic hearing was the mounting evidence of the decline in the share of federal R&D going to small businesses, as well as broader difficulties among small business in raising capital. At the same time, contemporary research suggested that small business were a fertile source of job creation further improved the climate for SBIR legislation.[17]

GOALS OF THE LEGISLATION

The legislation authorizing SBIR had two broad goals from the outset. According to the report language accompanying the legislation:

[14] Lerner, *op. cit.*

[15] *Ibid.*

[16] David B. Audretsch, *Innovation and Industry Evolution.* Cambridge, MA: The MIT Press, 1995.

[17] David L. Birch, "Who Creates Jobs?" *The Public Interest.* Vol. 65, 1981, pp. 3-14.

> "The purpose of the bill is twofold: to more effectively meet R&D needs brought on by the utilization of small innovative firms (which have been consistently shown to be the most prolific sources of new technologies) and to attract private capital to commercialize the results of Federal Research."[18]

More specifically, the 1982 act creating SBIR listed four program objectives:

1. To stimulate technological innovation
2. To use small business to meet Federal research and development needs
3. To foster and encourage participation by minority and disadvantaged persons in technological innovation
4. To increase private sector commercialization of innovations derived from federal research and development.

THE SBIR SET-ASIDE

To carry this out, the act required federal agencies with R&D budgets in excess of $100 million to set aside 0.2 percent of their funds for SBIR. This totaled $45 million in SBIR's first year, fiscal year 1983. Over its first six years of operation, the percentage set-aside increased to 1.25 percent. Modeled largely on the NSF's small business initiative, SBIR grants had three phases. Phase I, essentially a feasibility study, was a modest grant (in the neighborhood of $25,000 in 1983) to assess the technology's potential. Phase II, whose value was approximately seven times that of Phase I, was for more extensive development and prototype development. Phase III involved no federal funds, but was the stage in which award recipients were expected to receive private sector financing for commercialization.

Today, Phase I grants can be as high as $100,000 and Phase II grants are typically approximately $750,000. Across all agencies, SBIR's funding for FY 1998 totaled $1.1 billion, with the Defense Department having the largest program at $540 billion and the National Institutes of Health following with some $266 million in FY 1998.[19] Since the program's inception in 1983, SBIR has made over 45,000 awards, totaling $8.4 billion in 1998 dollars.[20]

---

[18] U.S. Senate, Committee on Small Business (1981), Senate Report 97-194, *Small Business Research Act of 1981,* September 25, 1981.

[19] See HtmlResAnchor http://www.acq.osd.mil/sadbu/sbir/overview.html for information on DoD's SBIR program. For information on NIH's SBIR program, see HtmlResAnchor http://grants.nih.gov/grants/funding/sbir.htm#sbir.

[20] For an overview of the origins and history of the SBIR program, see the recent article by James Turner and the late Representative George Brown, former Chairman of the House Committee on Science, "The Federal Role in Small Business Research," *Issues in Science and Technology.* Summer, 1999, pp. 51-58. This article provides a critical appraisal of the program, highlighting a number of issues raised in the symposium.

THE 1992 REAUTHORIZATION

Congress made two major changes to the SBIR program in the 1992 reauthorization. First, the set-aside rate doubled to 2.5 percent. Second, Congress increased the emphasis on commercialization of federal R&D as a program goal. Whereas the original SBIR legislation specified that Phase I grants should demonstrate "scientific and technical merit," the 1992 reauthorization stated that Phase I should demonstrate "scientific and technical merit and feasibility of ideas that appear to have commercial potential."[21]

With respect to Phase II, the 1992 SBIR legislation substantially increased the importance of commercial potential. In evaluating Phase II applications, agencies were directed to assess a technology's commercial potential as evidenced by:

1. The small business' record of commercializing SBIR or other research;
2. The existence of second phase funding commitments from private sector or non-SBIR funding sources;
3. The existence of third phase, follow-on commitments for the subject of the research, and;
4. The presence of other indicators of the commercial potential of the idea (GAO, 1999).

This legislative language reflected Congress' desire to encourage a higher commercialization rate for SBIR-funded technologies.

SBIR REAUTHORIZATION

The SBIR program is scheduled to expire on October 1, 2000 and the Congress held reauthorization hearings on the program in 1998 and 1999. Because of the size of SBIR, its emphasis on research and commercialization, its goals of meeting agency research needs while contributing to economic growth, the program represented an ideal starting point for fact-finding phase of the STEP Board's project on "Government-Industry Partnerships for the Development of New Technologies."[22] It is in this context that the Board on Science, Technology, and Economic Policy (STEP) organized a one-day symposium to review the program. The objective of the symposium was to review the history, rationale, operational issues, and current academic research regarding the program.

---

[21] For more information on the points raised in this paragraph, see Robert Archibald and David Finifter, "Evaluation of the DoD SBIR Program and the Fast Track Initiative: A Balanced Approach," in National Research Council, *The Small Business Innovation Research Program: The Fast Track Pilot.* Washington, DC: The National Academy Press, 1999.

[22] See the Preface for an explanation of the project's origins and objectives.

CRITICAL PERSPECTIVES ON SBIR

While the SBIR program generally enjoys bipartisan support, it is not free from criticism and controversy. First, SBIR is seen in some quarters as a tax on R&D funds. Given the demands for extramural R&D funding, some object to that 2.5 percent of this total must be allocated to the SBIR program. A second and related concern involves the different constituencies for R&D and small business. In the Congress, the science committees see themselves as the stewards of the public dimension of the nation's scientific and research enterprise. To some extent, SBIR can be seen as serving a different constituency, namely small business, although it can also be seen as a means to commercialize the results of publicly funded R&D. Some members of the university research community see SBIR as a drain on R&D resources that might be better utilized within a university environment and better allocated through traditional channels. In addition, some economists object to the SBIR program as an unwarranted and unnecessary intervention into capital markets.[23]

AN OVERVIEW OF THE SYMPOSIUM[24]

In opening the symposium, Mark Myers, vice president for research at Xerox Corporation, pointed to one virtue of the SBIR program, which is bringing small business into the family of companies doing business with the federal government. This not only promotes the growth of small businesses, an important source of innovation in the U.S. economy, but it also opens up the government to new ideas in the procurement process. At the same time, Dr. Myers noted that although the SBIR program has been around for a number of years, the close scrutiny of the day's program might well shed light on ways to improve SBIR.

Duncan Moore, associate director of the White House Office of Science and Technology Policy, framed his opening remarks both from his perspective as a public official and from his own experience as a one-time SBIR award winner. A firm he founded in 1980 when he was a university professor was unable to obtain venture capital or bank financing. Dr. Moore's firm turned to SBIR, which served a vital role in the technology's development. While very important to his firm, Dr. Moore also recalled how the gap between Phase I and Phase II funding was a problem; his firm, for example, had to turn to consulting and other work not central to its mission to weather the period of time between grants. From a personal perspective, Dr. Moore said that the issue of "SBIR mills"—firms that seem to exist only to win SBIR grants, but not commercialize technology—should

[23] See, for example, Scott Wallsten, "Rethinking the Small Business Innovation Research Program," in Branscomb and Keller, eds., *Investing in Innovation*. Cambridge, MA: The MIT Press, 1998, pp. 194-220. This view is not shared by other researchers; see, for example, Josh Lerner's paper in the Annex to this volume.

[24] This section summarizes the proceedings of the conference held on February 28, 1998 in Washington, D.C. at the National Academy of Sciences.

somehow be addressed. He suggested that firms be required to write business plans that outline how the technology will be commercialized as a condition for receiving an SBIR award. Dr. Moore also said that, while some university professors see SBIR as a tax on R&D funds that reduces research funding for them, academicians should see SBIR as an opportunity for them to spin-off their research into companies.

History and Legislative Perspective on SBIR

In recalling the early years of efforts to direct federal R&D funds to small business, Roland Tibbetts said that as early as the mid-1970s, many members of Congress understood that small businesses were a potentially fruitful source of innovation. However, government agencies were wary of funding high-risk technologies that showed commercial promise. In fact, using the term "commercial potential" in an SBIR application was likely to do more harm than good for the applicant. A desire to encourage commercialization has emerged among many agencies in recent years, and Mr. Tibbetts characterized this as a promising development. Referring to his own research, Mr. Tibbetts affirmed his belief that the NSF SBIR program has been successful in generating innovations and job creation among small firms.

In commenting on Mr. Tibbetts remarks, Paul Cooksey and Patricia Forbes, majority and minority staff directors respectively of the Senate Committee on Small Business, noted the widespread popularity of the program. Ms. Forbes reminded participants that SBIR successes could be found not just in high-technology states such as California and Massachusetts, but across a wide range of states. Mr. Cooksey commented on the bipartisan support that SBIR enjoys, but noted that because SBIR is not a high profile program, its supporters must be vigilant in cultivating support and inviting enough scrutiny so that SBIR is not seen as "stealth" program.

From the perspective of the House Committee on Science, James Turner said that even though SBIR is a dispersed program, it is substantial in size. At $1.2 billion, it should be subject to the same level of scrutiny as any large program. He also noted the "schizophrenic nature" of the SBIR program—namely the program's charge to support agency research needs while at the same time promoting private commercialization—suggesting that was one area which might usefully be explored in deliberations on SBIR. Mr. Turner encouraged rigorous assessment of SBIR, and he concluded by saying that the House Committee on Science, as advocate of research and competitiveness, could contribute to strengthening the program.

Research Perspectives on SBIR

In introducing his research, Joshua Lerner of the Harvard Business School concurred with Mr. Turner's view that SBIR has received little attention from the research community, given the program's size. Dr. Lerner said that he found that SBIR awardees generally grew at a greater rate than comparable non-SBIR firms—whether measured by employment growth or job growth. The greatest job growth is associated with regions of the United States, such as California or Massachusetts, with significant high-technology entrepreneurial activity. The picture that emerges from the research is a program that is working effectively and appears to be playing a positive role in stimulating small-firm creation. Dr. Lerner said that an SBIR award appears to have a "certification" function, acting as a stamp of approval for young firms to obtain resources from outside investors.

From a policy perspective, Dr. Lerner said that the rationale for this type of "public venture capital" program rested on two economic arguments. First, there are significant knowledge spillovers associated with R&D, that is, the benefits of R&D do not accrue only to those making the R&D investment. Second, information problems hamper investors' efforts to identify promising technologies. The large number of companies seeking financing, coupled with the technological sophistication and business uncertainty involving innovative business proposals, pose significant challenges for potential investors, challenges that may result in underinvestment in new technologies. This problem may be especially acute for small firms.

Case Studies

While aggregate measures of the SBIR's impact are valuable, the next panel reviewed the experiences of firms that have received grants, bringing together venture capitalists, award winners, and government officials. This panel provided useful insights into the program's contributions to small business growth, as well as ways in which the program could be improved. Gary Morgenthaler, now a venture capitalist, described a software firm he founded on the strength of three SBIR awards, totaling $650,000, that eventually grew to a company with $160 million in revenue and 1,300 employees. Dr. Morgenthaler observed that it is important to know how to win grants, but that the best way to win several SBIR awards is to achieve promised goals. As a venture capitalist, Dr. Morgenthaler noted the difference between how SBIR operates and how venture capitalists evaluate opportunities. The SBIR program places a greater emphasis on good science in evaluating applications, whereas a venture capitalist looks to commercial potential and supplies management and marketing expertise to companies that receive venture funding. Echoing Dr. Moore, Dr. Morgenthaler also addressed the funding gap between Phase I and Phase II grants and the issue of "SBIR mills." Dr. Morgenthaler said that the 18 month period between the initial application and Phase II is "an eternity" in today's market. With respect to SBIR mills, Dr. Morgenthaler said that the existence of firms that receive between 50 and 300 awards, with modest commercialization records, may not be the best use of taxpayer funds. However, he did suggest that closer collaboration between the venture capital industry and the SBIR program might improve the chance that good research will be coupled with effective management and adequate funding.

Richard Carroll, of Digital System Resources, described how a small business such that establishes a relationship with an agency through SBIR can contribute innovative ideas to the agency as the business grows. Digital System Resources, which developed a significant new SONAR technology for the U.S. submarine fleet, uses "innovation focus groups" among its 200 employees to develop new ideas to meet the needs of the Defense Department. Using SBIR, the Navy has procured innovative technology from his company, and Digital Systems has had a steady customer for its technology as the firm has grown.

Operational Challenges

In discussing program administration and issues in SBIR's rationale, Dan Hill of the Small Business Administration (SBA) recalled a discussion that took place at SBIR's 1982 authorization hearing. An NSF official said that creating SBIR would amount to establishing a new set-aside program. In response, Senator Warren Rudman asked how many NSF grants went to small businesses and the NSF official said that none did. Senator Rudman then responded that SBIR did not constitute the creation of a new set-aside program, but the destruction of an old one. In this context, Mr. Hill said that SBIR should not be viewed as a 2.5 percent tax, but should instead be compared with the 97.5 percent of federal R&D conducted outside the SBIR program. As the SBA assistant administrator for technology, Mr. Hill identified the following operational challenges for SBIR as the program went through its upcoming reauthorization:

- **Commercialization**: the extent to which commercial potential should be weighed in the award process.
- **Cost-sharing**: the extent which agencies should decide awards on the basis of the amount cost-sharing provided by the applicant.
- **Multiple award winners**: how to address the SBIR awardees that win many awards, but whose commercialization record is modest. Mr. Hill noted that agencies have sufficient authority under current legislation to limit multiple awards.
- **Time delays**: whether too much time passes between Phase I and Phase II awards.
- **Geographic distribution**: whether anything should be done to address the fact that awards tend to be clustered in specific regions.
- **Evaluative criteria**: given that SBIR is administered differently in different agencies, Mr. Hill said that developing consistent and improved evaluative criteria was necessary for SBIR to comply with the Government Performance and Results Act.

- **Set-aside percentage**: Mr. Hill said that SBA would like to see the percentage rise, but preferred to see the overall federal R&D budget increase.

In taking up some of these themes, Dr. Carl Nelson, formerly with the Ballistic Missile Defense Organization (BMDO), suggested two ways to improve SBIR. First, to permit government decision makers to better respond to the rapidly changing marketplace, Dr. Nelson suggested creating incentives, such as a $10,000 bonus, to SBIR program managers whose awards perform well. Second, to address multiple award winners, Dr. Nelson recommended a formula by which a ceiling would be put on the percentage of an agency's SBIR budget that could go to a set of firms whose SBIR awards had attained a specified threshold.

David O'Hara of Parallax Research raised the issue of cultural differences across different firms that participated in SBIR. At a firm for which he worked prior to Parallax, Mr. O'Hara recalled that his boss there was interested in winning SBIR awards, but uneasy with a provision in BMDO's SBIR program requiring evidence of commercial potential before a Phase II grant was awarded. Mr. O'Hara, who was interested in bringing products to marker, realized he was in the wrong corporate culture, and left to form Parallax. Based on this experience, Mr. O'Hara encouraged policymakers to look for ways to attract firms to SBIR that were actively interested in commercialization.

In reflecting on the comments of Dr. Nelson and Mr. O'Hara, Gene Banucci of Advanced Technology Materials, Inc. (ATMI) said that the pressure from BMDO to have ATMI obtain third-party financing as a condition for ATMI's Phase II award proved very helpful to his company. The pressure reinforced a trend begun within ATMI to look for partners to grow the company. This strategy, using SBIR funds and other partners, has enabled ATMI to grow to $102 million in revenues and eventually graduate from SBIR because, with more than 500 employees, ATMI no longer qualified as a small business. ATMI has been a commercial success with its technology to safely transport toxic and hazardous gases used in semiconductor fabrication, while also meeting BMDO's need for such technology. In characterizing SBIR as "a terrific program," Dr. Banucci said it could be further improved by increasing the size of awards to firms who have lined up several partners to share costs.

Improving Assessment and Selection

To draw lessons on how to improve SBIR selection, assessment, and administration, the symposium heard from representatives of several agencies' SBIR programs and a non-SBIR technology investment program, the Advanced Technology Program (ATP). The ATP differs from SBIR in that it is targeted at technologies with commercial potential from the outset, and at technologies with have spillover potential, that is, high-risk technologies with widespread economic impact. ATP's evaluation program is very detailed and developed, and the ATP's Dr. Maryellen Kelley described the data collection process employed by ATP to conduct evaluations. Speaking for the National Institutes of Health, Dr. Herbert Kreitman noted that approximately 95 percent of NIH SBIR awards were research grants, so that commercialization receives less attention at NIH than elsewhere. NIH is also very interested in investigator-initiated research, so that NIH is unlikely to publish solicitations in specific technology areas. Nonetheless, he stated that the General Accounting Office has found NIH's commercialization record in SBIR to be very good. Dr. Kreitman concluded that NIH could further improve its commercialization record if it had more funds for outreach to the community of small businesses who do, or might do, research with NIH.

From the DoD perspective, Jon Baron, the DoD SBIR program manager, described efforts to improve commercialization through the Fast Track pilot program. Although aware of a number of very successful commercial products generated by DoD's SBIR program, the Department wanted to improve the rate of commercialization. Through Fast Track, DoD hopes to encourage commercialization by reducing the time period between Phase I and Phase II awards, as long as firms show that they have commitments from third parties to buy the SBIR technology. Mr. Baron said that early indications were that Fast Track has enabled SBIR companies to leverage the SBIR grant, that is, the promise of expedited Phase II funding has enticed some private investors to provide matching funds to the SBIR applicant.

In commenting on Dr. Kreitman's and Mr. Baron's remarks, Charles Rowe of the House Committee on Small Business said he was very pleased to see Fast Track programs at NIH and DoD. The funding gap between Phase I and Phase II is one of the most pervasive complaints among small businesses participating in SBIR. To see policy experiments underway at two agencies to address these complaints was heartening.

Perspectives of SBIR Program Managers

Having heard a variety of perspectives and suggestions about the SBIR program, SBIR program managers from different agencies provided their views on how the program operates and how it could be improved. Some SBIR program managers, such as the Energy Department's Arlene de Blanc saw themselves as "matchmakers"—bringing the appropriate small business to bear on specific agency mission needs, in a way that encourage marketplace success. Being a successful matchmaker can be a delicate balancing act, and several program managers remarked on the need for flexibility in program administration to allow matchmaking function to be carried out successfully. The U.S. Army's Kenneth Gabriel suggested that with entrepreneurship taking on even more importance in the U.S. economy in terms of job creation and innovation, it was even more important to use the SBIR program to seek out creative elements in the economy to achieve government missions. Dr. Gabriel stressed that SBIR managers should continue to adhere to the highest standards of technical merit, but he argued that the program must be decentralized in administration so that program managers can encourage risk-taking where appropriate.

Such decentralization may mean that different agencies may define "SBIR success" differently. The Defense Department may see success as a piece of hardware that goes into a weapons system, whereas NIH may view success as an article published in scientific journal that stimulates additional inquiry. Dr. Robert Norwood of NASA expressed his concern that any one measure of success could apply to a program administered through eleven different agencies. Dr. Norwood also cautioned against relying on individual success stories in evaluating the program as a whole. Among program managers, the recent emphasis on commercialization was generally regarded as a positive development, but they cautioned that Congress should not lose sight of mission needs or of the limited resources within agencies to promote commercial goals of SBIR.

CONCLUSION

In summarizing the day's discussion, Mark Myers observed that the SBIR program appears to enjoy broad-based bipartisan political support. Dr. Myers noted that SBIR's focus on technical merit and its competitive award process are time-proven approaches to ensure quality. Yet despite its substantial budget, this decentralized program has been subject to relatively little analysis. The multiple goals of the SBIR program evaluation present a challenge, especially when even the definition of commercialization is open to different interpretations by different agencies. Dr, Myers emphasized that the participating agencies' autonomy is such that there are, on a *de facto* basis, many different programs operating.

Dr. Myers also endorsed the Defense Department's recent emphasis on greater speed in program operations, remarking that improving the pace of decision-making is an important goal from the perspective of the private sector. With the quickening pace of innovation, it is important to improve government decision-making cycles; as Dr. Myers observed, "everything you can do to speed up anything" is considered a virtue in the private sector. If an objective of SBIR is to use the private sector to meet government mission needs while encouraging commercialization, then exploring ways to make SBIR program administration fully responsive to the private sector is important. Dr. Myers noted that another advantage to the program is that it makes available an alternative source of very patient capital, which can contribute to sustained support for research and development. In closing, Dr. Myers thanked attendees for their participation in a symposium that generated rich and valuable discussion on SBIR. He expressed his confidence that the symposium would be a significant first step in the STEP Board's efforts to improve our understanding of the history, goals, and operations of the SBIR program, as well as the challenges and opportunities it faces.

III

PROCEEDINGS

Welcome[1]

Charles W. Wessner
National Research Council

Welcoming the symposium participants, Dr. Wessner introduced the day's activities as the first element in a program examining "Government-Industry Partnerships for the Development of New Technologies." Being carried out under the aegis of the National Research Council's Board on Science, Technology and Economic Policy (STEP Board), direct responsibility for the meeting effort rests with a distinguished steering committee chaired by Gordon Moore, chairman emeritus of Intel Corporation; its membership includes Mark Myers, senior vice president of Xerox Corporation, leading our discussion today; William Spencer, chairman of SEMATECH; Gordon Binder, chairman of Amgen; Kenneth Flamm, senior fellow at The Brookings Institution and a speaker at the symposium; and other distinguished academics and industrialists.

The basic objective of the Government-Industry Partnerships project, as characterized by Dr. Wessner, is to determine "best practice" in government-industry partnerships taking into account sectoral differences with respect to issues such as project selection, private-sector management, balanced technical contributions among participants, effective intellectual property protection, and policies to encourage the commercial exploitation of results. Neither the Government-Industry Partnerships project nor today's symposium would "engage in ideological discussions of whether the government should do what it has in fact done for the last 200 years," but would instead seek to review accomplishments, identify problems, and make specific, pragmatic suggestions.

[1] This section summarizes the conference held at The National Academy of Sciences on February 28, 1998.

In addition to the issue of best practice, the project will examine:

- the rationale and national benefits to be derived from government support to collaborative efforts by industry to bring new technologies to market;
- the principles that should guide such cooperation, demarcating the role and contribution of the public authorities;
- the different types of cooperative programs, including the rationale for strategic alliances among firms in sectors that are supported by publicly funded programs; and
- the sources of past successes and failures of government–industry cooperative programs designed to support the development of U.S. industries based on promising new technologies.

Also among the objectives of the project as a whole are:

- promoting cross-fertilization among U.S. government programs, which Dr. Wessner described as "very rare";
- benchmarking U.S. practices against foreign practices, "not necessarily to emulate foreign practices but to understand on one hand, what the competition is doing"; and on the other, identify opportunities for mutually beneficial cooperation;
- arriving at recommendations for improved American policymaking on national and international partnerships.

Several interesting trends are behind the interest and growth in government partnerships. The cost of developing new technologies, the risks associated with such development, the dispersal of technological expertise, the corresponding "technological merge," and the difficulty for any one company to capture all the benefits, all point to the need for cooperation between industry and government at its various levels as the world enters the twenty-first century. To underscore the importance of government–industry collaboration, Dr. Wessner challenged the participants to "name one major, export-oriented industry in the United States that has not received very substantial assistance through its growth period from the government." Although it takes many forms—some direct, some indirect—support for industry has pervaded the economy, especially in high-growth, high-technology sectors as semiconductors, computers, telecommunications, and aircraft. American markets are among the world's most dynamic, with levels of competition and innovation that have proved to be the source of remarkable economic growth. Recognizing the important, sometimes decisive role the government plays in devising a policy framework that supports innovation and sustains cooperation is essential for our continued success.

Introducing Dr. Myers, a member of the STEP Board, Dr. Wessner described him as "responsible for some of the best technology coming out of American corporations today" as vice president for Corporate Research and Technology at Xerox. Recalling a recent visit to the research facility Xerox PARC in Palo Alto, Calif., Dr. Wessner said that he had left "impressed and encouraged about what the unique blend of American capitalism and supportive government policies has to offer" the next generation of Americas and the world.

Symposium Introduction

Mark Myers
Xerox Corporation

Speaking on behalf of the STEP Board, Dr. Myers welcomed the participants to the symposium on the Small Business Innovation Research (SBIR) program, extending particular recognition to White House representatives Dorothy Robyn of the National Economic Council and Duncan Moore of the Office of Science and Technology Policy (OSTP); Paul J. Hoeper of the U.S. Department of Defense (DoD); Martyn Evans, Shadow Minister for Science and Technology from Australia; and officials from other U.S. government agencies and the diplomatic corps. Noting the very strong attendance, Dr. Myers remarked that we could legitimately infer policy interest in the SBIR program as the source of "one of the best turnouts" for a STEP Board event.

The Need to Review SBIR

Introducing the day's discussions, Dr. Myers stated that the symposium would review the types of benefits that the SBIR program "brings to small business in the United States in particular and [to] the American economy in general," which is in line with a perceived desire "to make such programs effective so that they can become even more useful in the future." Previewing the day's program, he listed the following as among its goals:

- reviewing the SBIR program's history and rationale, including some "common understandings and misunderstandings" surrounding it. In this connection, he expressed gratitude for the participation in the panel on "History and Current Legislative Perspective on the SBIR Program" of

Roland Tibbetts, former Program Manager of the SBIR program at the National Science Foundation;

- reviewing the "best research that's been done" on SBIR. He pointed to the presence of Josh Lerner of Harvard Business School who would be presenting his recent analysis of the SBIR program under the heading of "Research Perspectives on the SBIR"; and
- examining successful outcomes of SBIR awards, with a focus on elements that have contributed to their success and that might contribute to the success of future awards. He thanked Dr. Robyn for agreeing to moderate the symposium's panel on "Case Studies" designed to bring the views of industry directly into the discussion.

Identifying Challenges to SBIR's Effectiveness

Dr. Myers emphasized, however, that the symposium would not limit its scope to the accomplishments of the SBIR program, which he said have been many, but also would identify challenges it faces. "Clearly, no program is perfect," he stated. "Indeed, coming from the private sector, I can assure you that perfection—that is, always getting it right-exists neither in the public nor the private sector." The key in the private sector is to do better than the competition over time in serving one's customers, he noted, whereas the key in the public sector is "to serve the citizens and to ensure worthwhile return on investments . . . and to constantly find ways to improve them." Rep. Tom Davis (R-Va.), who serves on the House Committee on Science and the House Committee on Transportation, would offer thoughts on how to improve the SBIR program, as well as on its importance in furthering innovation in the U.S. economy.

Dr. Myers pointed to the presence of such "expert practitioners" as Carl Nelson, formerly of the Ballistic Missile Defense Organization, on the panel titled "Program Challenges—Operational Views," to be led by Dan Hill, the Small Business Administration's Assistant Administrator for Technology.

He expressed his particular enthusiasm, as "someone who is responsible for research and development (R&D) in a large corporation," for the panel on what he called the "thorny questions" of "Improving Assessment and Selection," to be moderated by Dan Roos of the Massachusetts Institute of Technology. Dr. Myers observed that some might wonder why a discussion of the Advanced Technology Program (ATP), which is managed by the National Institute of Standards and Technology, is included in a program on SBIR. The answer is that one of the objectives of the Government-Industry Partnerships project is to learn from the experience of different U.S. and foreign technology development programs. "This cross-fertilization, we believe, will prove quite valuable," he said.

Similarly, the day's program would draw on the experience of two of the largest SBIR programs, those of the National Institutes of Health and those of the Department of Defense, which some believe have significantly different objectives and measures of accomplishment. Their comparison is intended to lead the panel to suggest ways of improving understanding of the different agency requirements, the metrics for success, and the overall performance of the SBIR program. To have a down-to-earth reaction to the day's discussion, the final panel, "Observations and Policy Issues: Agency Perspectives," includes program managers from some of the largest SBIR programs.

Pointing out that "some of the best experts in the country on the development of new technologies in a small business, entrepreneurial environment" were in attendance, Dr. Myers said that he looked forward to a "rich exchange of views." He then introduced Dr. Moore, the Associate Director of Technology at the White House Office of Science and Technology Policy, noting that the two have known one another for many years as fellow residents of Rochester. Dr. Moore's career includes service as professor of optical engineering at the University of Rochester, as director of the Institute of Optics and dean of engineering and applied sciences at that institution, and as founder and president of Gradient Lens Corporation of Rochester.

Opening Remarks

Duncan Moore
White House Office of Science and Technology Policy

The Contribution of High-Tech Firms to the Economy

Dr. Moore, introducing himself as a winner of SBIR awards from both the NSF and the DoD, stated that he would talk about his personal experiences with the program. First, however, he wished to put forward the contention that the U.S. technical community has not taken enough credit for the reduction of the federal budget deficit. This community has invested in R&D for the past 50 years, developing the engine that has created the country's economic growth. This investment has contributed to a large number of jobs, particularly in the last decade, that has increased the amount of wages paid and, therefore, taxes collected.

He pointed to an unemployment rate that he said was at "almost an all-time low"; significantly, although the national jobless rate is at about 4.7 percent, it is at about 1.9 percent for those with college degrees. Vigorous demand for university-trained workers is contributing to the continued growth of the American economy.

He noted the fading of the conventional wisdom of a half-decade ago that "the United States was great at basic science—we won the Nobel prizes—but Japan was much better at taking these ideas to product." This has changed. Although economic difficulties in Japan may account for part of the change, another reason is that U.S. programs have been created at the federal and, in particular, the state level to address the questions: "How do we get the ideas out of the universities into the marketplace?" and "How do we create jobs locally?" Reminding the audience that economic development is in the end always "a local . . . not a federal game," he pointed to the political imperative faced by mayors, county commissioners, and state legislators to create jobs.

Another lesson from recent experience is that, given the diversity of needs among both geographical and industrial sectors, no one program is appropriate for all circumstances. For example, policymakers have recognized that "biotech is very different from automobiles," yet both may face challenges requiring partnering.[1] The emphasis on spin-offs, whether of jobs or companies, has evolved into emphasis on partnering. "The implication of 'spinoff' is, that an idea is conceived and it is hoped that a company will develop from that idea," Dr. Moore explained, "whereas 'partnership' implies that the partnership is going to be there for a longer period of time than it would in the case of when an idea is transferred to a new company."

He pointed to such programs as SBIR, ATP, NSF's Experimental Program to Stimulate Research, and the Experimental Program to Stimulate Technology, the latter being designed to help the 17 states that have attracted lower-than-average amounts of research funding. Initial skepticism on the part of Dr. Moore, as a resident of New York State, toward programs aimed at helping smaller states has given way to strong support. The belief that "economic survival is based on having good R&D and having good universities in your home state," as well as the desire "that all regions of the country move forward," must be backed, he said, by "processes" allowing states with fewer resources "to move forward in these wonderful economic times." He noted that SBIR's total annual budget represents a "relatively small number of dollars as a percent of total R&D": At about $1.1 billion, it amounts to between 2 and 3 percent of the federal R&D portfolio, depending on how the portfolio is calculated.

Personal Experience with SBIR

Dr. Moore then summarized the experience with SBIR awards he had gained through his own company, founded in 1980. He recalled turning to the program in the technology development stage, after finding neither venture capitalists nor banks willing to back a firm without a product. A New York State program that offered funding to help winners of Phase I SBIR grants bridge the gap to Phase II made the SBIR program more attractive to Dr. Moore; his firm submitted two SBIR proposals to the NSF, and both eventually won funding.

The Phase I – Phase II Gap

Nonetheless, to stay in business, the "huge" gap between Phase I and Phase II caused his firm to take consulting jobs and other work that were not part of the

---

[1] For a review of one such program, see *Review of the Research Program of the Partnership for a New Generation of Vehicles (PNGV): Fifth Report*, National Academy Press, Washington, D.C., 1999.

firm's main mission. There were other difficulties: Corporate and personnel changes disrupted agreements the firm had made to work with large corporations on Phase III of both of its SBIR projects, and a shift in the market undermined a third arrangement.

Dr. Moore called on his experience to compare the NSF's SBIR program with that of the DoD, to which his firm turned at a later stage. From the comments he received on his proposals to NSF, he concluded that the ideas the reviewers favored were "further from the product . . . They're looking at the five-, ten-year ideas, not the ones that are going to be products 18 months from then." He characterized DoD's SBIR program as "very directed" in comparison: His firm pursued work in one of the DoD program's categories, and the DoD would have been guaranteed as a customer had the Cold War not ended. Dr. Moore's technology finally yielded a product in 1995, 26 years after he had begun research on it.

Key SBIR Issues

Dr. Moore then highlighted a number of issues often discussed in relation to the SBIR program:

- SBIR mills: Some companies are alleged to exist entirely for the purpose of winning grants while lacking any serious intent to commercialize. Speaking personally rather than on behalf of the White House, he advocated that firms should first write their business plans, then indicate how SBIR-backed research might fit with the plans;
- R&D tax credits versus SBIR: Dr. Moore noted that a tax credit would have been of no value to his firm, which was not generating profit and "needed the cash flow" to make payroll; and
- Universities versus SBIR: Opposition from academics who see money that goes to SBIR as reducing funds available for their research was characterized by Dr. Moore as "very short sighted." SBIR grants offer professors and students an "ideal" opportunity for spinning off a company, and the university community should be looking at the program in a positive light "as part of economic development."[2]
- Employment Generation: Dr. Moore also stressed the employment generation that is derived from the small business community. For example, Small Business Administration figures show that companies with fewer than 500 employees contributed 10.5 million jobs to the U.S. economy from 1991 to 1995. In the same period, employment in companies with

[2] See Audretsch, David B. et al., "Does the SBIR Foster Entrepreneurial Behavior? Evidence from Indiana." In *The Small Business Innovation Research Program: Assessing DoD's Fast Track Initiative,* National Academy Press, 1999.

more than 500 employees fell by 3.2 million. Underscoring the Clinton administration's support for the SBIR program, Dr. Moore welcomed this effort to review the program, noting that "every program can benefit from continuous improvement," a principle he called "part of the business climate today."

Panel I

History and Current Legislative Perspective on the SBIR Program

INTRODUCTION

Robert Neal
Office of Small and Disadvantaged Business Utilization
U.S. Department of Defense

Introducing the day's first panel on the origins and history of SBIR, Mr. Neal stated that the office he directs spends a great deal of time supporting the program and derives considerable benefit from it. He recognized the panel's main presenter, Roland Tibbetts, as being able to speak from a position of long and extensive experience with the program.

ORIGINS OF AND COMMON MYTHS ABOUT THE SBIR PROGRAM

Roland Tibbetts
National Science Foundation (ret.)

A Perspective on SBIR's Evolution

Declaring his delight that the SBIR program is "on the radar screen of new public policy," Mr. Tibbetts expressed the hope that those in attendance would take an active part in enhancing its role and visibility. He recalled that, when he was a director of administration for NSF's former Research Addressing National Needs program in the Nixon administration, the small-business community approached the Congress out of dissatisfaction over the fact that, despite numerous requests, the level of federal R&D money it received had been stuck at 3.5 percent for at least 15 years. Congress then directed NSF, over the agency's opposition, to place 7.5 percent of the RANN program's funds for 1976 with small business firms.

Because of the sensitivities occasioned by the Golden Fleece Awards that Senator William Proxmire (D-Wis.) was giving out at the time to stigmatize wasteful government spending, Mr. Tibbetts was asked to come up with a program that would ensure that the proposals submitted by the small firms would be of high quality.

A Changing R&D Landscape

The U.S. R&D landscape was very different 20-25 years ago: Small, high-technology firms had been concentrating in Silicon Valley and around Boston's Rt. 128 for some 20 years, but, although these areas were active then, they were "nothing like they are today." Detroit, meanwhile, was having problems with inflation, and unemployment in the auto industry was high.

Even at that time, Congress understood "that small firms were good at technological innovation, and that technological innovation was important to economic growth." Commissions formed as long ago as the mid-1960s to look at the importance of small, high-technology firms that had repeatedly issued reports and recommendations highly favorable to small business. Still, very few of those recommendations were acted upon, because, Mr. Tibbetts recalled, "they were opposed by other recipients of federal R&D" funding.

Today's landscape provides a sharp contrast:

- The eastern half of Massachusetts and in the area south of Sacramento, California are "alive with small, high-tech firms."
- Venture-capital investments reached new highs in 1996 and 1997, rising 20 percent in 1996.[1]
- Initial public offerings of stock were at record levels in 1996 and fell just short of a new record in 1997.
- Small Business Investment Corporations are currently at record levels of investment.
- Of those scientists and engineers primarily engaged in federal R&D, the number working for firms with 500 or fewer employees exceeds the number at universities. The number of scientists and engineers employed by small firms is approximately one-fifth of all scientists and engineers.
- The importance of high-technology firms is generally recognized by the

[1] For a brief review of venture captial in the U.S., see the paper by Harvard Business School's Joshua Lerner in the Annex.

public—but, even including the awards that SBIR distributes, small firms continue to receive only about 5 percent of federal R&D funding.

SBIR's Early Years

Mr. Tibbetts cited numerous reasons for his interest in small, high-tech firms beyond his desire to increase their participation in government R&D. As vice president of two such firms for about 12 years and as founder and director of Allied Capital in Washington, he had observed that the cohort of executives of such firms was "possibly the best" that he had ever seen: "You are smart if you are in high technology; you are extremely courageous; you are extremely hard working; you are willing to risk a great deal—including your job, because you are confident you can get another one; [and] you are willing to tackle—in fact, you want to tackle—breakthrough ideas, which are tremendously important to the future of the country." But despite the great and growing importance of small, high-tech firms to the economy, "they still have a distinct problem" in winning recognition in the sphere of R&D policy.

Efforts by Senator Edward Kennedy (D-Mass.) in 1976 helped raise small firms' share of NSF's R&D funds from 7.5 to 10 percent, and the following year an SBIR program came on stream at NSF. NSF made 42 Phase I awards in the program's first year, and about the same number of awards in each of the two subsequent years. Enthused by the program, small firms approached other federal agencies to propose that they follow NSF's lead; after meeting with a cool reception in most cases, the firms went to Congress and the Small Business Administration, with the result that legislation was passed in 1982 creating a new program, which started up in 1983. To date, the SBIR program has made 40,000 awards worth $7 billion to 9,000 firms.

When the program began, Mr. Tibbetts recalled, "it was almost impossible, with a high-risk idea, to get funding, and almost equally difficult to get an acceptance from a venture-capital firm" or from the individual "angel" investors that could then be found.

Additionally, he was "appalled" to find that "there was no interest in the government in maximizing return on investment," and that using the term "commercial potential" in a proposal was likely to do more harm than good. He hoped that changes could be brought about leading to the recognition of small, high-tech firms' importance in converting government R&D into public benefit through technological innovation and commercial applications and in stimulating economic growth.

SBIR Program Design

Once recognition had been attained, Mr. Tibbetts asserted, the "trick" was to design a program that would effectively take advantage of what small firms had to offer: The creation of a three-phase format has been SBIR's answer to that challenge. The intention behind Phase I, which originally provided $25,000 over six months, was to "use the least amount of money for the shortest period of time" so that the largest possible number of ideas could be funded "to see whether they had promise." Intended to fund high-risk projects from idea to prototype, SBIR designed Phase I so that money would be saved for larger investments in Phase II; this allowed the program to "provide what amounted to preventure capital" to help a firm to lower its risk and put itself in a position to obtain private financing. "Most breakthrough ideas" Mr. Tibbetts told the audience, "are simply too high risk even for risk-taking venture investors. Consequently, if government, which has a great deal of investment in cutting-edge research, could allow small, high-tech firms to participate in government research, we would possibly lower the risk with successful research in areas that were acceptable to private investors. And I think this has happened."

The evaluation process under Phase I, whose base has risen to $100,000 today, has successfully allowed program managers to establish the validity of firms' ideas. At the same time, it has provided proposers the "enormous opportunity" of having perhaps six to ten engineers who are versed in the field evaluate their submissions.

Economic Impact of SBIR

Before leaving NSF, Mr. Tibbetts reviewed 50 awards granted between 1977 and 1992. The awards selected for review were those whose winners attributed at least $2 million in sales to the award; overall, the 50 firms pointed to $9.1 billion in direct and indirect sales they believed would not have accrued if the program had not existed. The $9.1 billion figure is 30 times NSF's total investment in SBIR from the beginning of the program; the associated private investment was $963 million, three times the total cost of NSF's program. Falling outside the realm of Mr. Tibbetts' review were many other awards that resulted in commercialization.[2]

Job Creation: Although acknowledging the difficulty of pinning down the number of new jobs resulting from the SBIR awards, "because people work on many different projects," Mr. Tibbetts noted that the 50 companies totaled 1,254 employees—and average of 25 employees per firm—at time of award, and that they created a total of 10,267 new jobs in the 15-year period on which the review was based.

Patents and Collaboration: The firms received a total of 1,109 patents in that period; they formed 394 research collaborations with universities, 404 with

[2] This paper, which was distributed at the meeting, is included in the Annex.

industrial companies, 111 with national laboratories, and 50 with other organizations.

Funds for New Ideas: All 50 firms said the major benefit of the SBIR program was that it was willing to fund ideas that they were unable to get funded before, and 45 of the 50 said that SBIR was critical to their growth, to start up, or to survival.

Indirect Benefits: The indirect benefits of the program, the firms said, were the ability to hire high-quality people, plus the credibility, which they would not have had without the program, that allowed them to attract better consultants, university collaborations, venture capitalists, and customers.

Mr. Tibbetts observed that SBIR was initially accused by "some groups, societies, and associations" of funding commercial research with public money. "An economist at NSF at the beginning of the program said, 'If you do one thing, make sure you do not substitute public funds for private funds.' And we did not. All SBIR funding is to pursue R&D on government agency needs. We did add something important, though: In Phase III we would pursue commercial applications to increase the return on investment from government R&D."

DISCUSSANTS

Patricia R. Forbes
Senate Committee on Small Business

Political Climate for SBIR

Ms. Forbes began by pointing to the popularity that the SBIR program enjoys among Democrats on the Senate Small Business Committee—the committee's ranking minority member Sen. John Kerry (D-Mass.), is a strong supporter—and by expressing her belief that Republicans on the panel back it as well. Calling the SBIR program "easy to like," she noted that since its inception it has extended more than 41,000 contracts and grants, representing more than $6.5 billion in research opportunities overall, to small, high-technology companies throughout the United States. Agencies that participate in the projects and small firms alike have shared Congress' enthusiasm for the program; since 1989 the SBIR program has received eight favorable evaluations from the U.S. General Accounting Office (GAO) and others.

She praised the program for serving "an invaluable function": fostering small businesses' participation in meeting federal R&D needs and helping draw private capital investment to their innovative work. Without the program, many technological innovations would not have been developed; there would have been no other way for small companies to finance the research or attract private capital at later stages of development.

A demanding competitive awards process ensures SBIR's rigor, but the program is nonetheless flexible, Ms. Forbes observed. Noting that coming up with a viable product or process may require numerous tries, and that lessons learned from several research products may only later result in marketable products, she labeled the program's ability to accommodate the innovation process "a strength, not a weakness." SBIR's program design, which permits agencies to decide whether their mission needs are best met by contracts or grants, has yielded high-quality products and processes that have been useful to the agencies, have often resulted in private-sector sales, and have bolstered international competitiveness of the United States.

Although she highlighted the commercial success of a SBIR award-winning firm from Massachusetts, Ms. Forbes stressed that cases of companies "located in recognized high-tech states . . . are not the whole SBIR program." Often overlooked have been the achievements of SBIR winners in parts of the country "not known for being technological or innovative centers." She underlined the cumulative economic and technological effect of such SBIR-winning firms.

Prior to the run-up to the legislative reauthorization of the SBIR program in 2000, the Senate Committee on Small Business will be looking for ways to support and strengthen the program. Ms. Forbes said that she planned to report the views and concerns of symposium participants to Senator Kerry and other Democrats on the committee.

DISCUSSANT

Paul Cooksey
Senate Committee on Small Business

Personal Perspectives on SBIR

Remarking on the presence at the symposium of people associated with the early years of SBIR, including Lee Mercer, a former aide to Senator Warren Rudman (R-N.H.) and now head of the National Association of Small Business Investment Companies, Mr. Cooksey regretted the absence of Tom Powers of the House Committee on Small Business, who had played a very influential role in developing the program.

Mr. Cooksey was introduced to the SBIR program not long after his arrival at the Small Business Administration (SBA) in 1991. At a meeting called to plan the agenda for a management retreat, Mr. Cooksey facetiously suggested having the head of the SBIR program at the agency speak so that he himself could learn about the program. "I looked around the room, there were the top 40 people at SBA, and none of them knew what the program was about either," he recalled.

Mr. Cooksey has since concluded that SBIR is "an exceptional program to target a very small amount of money from our government to some exceptionally fine and very effective, small, high-tech firms." Stressing how low the program's funding level of $1.1 billion per year is in the context of the overall federal budget, he contrasted it with the contribution that small business has made to job creation in recent years and commented that "the SBIR program creates the type of energy and innovation that is so essential in a small business community."

The Current Legislative Climate

With the SBIR program set to expire on October 1, 2000, the Senate Committee on Small Business is beginning to examine it, because both the committee's chairman, Senator Christopher Bond (R-Mo.) and Senator Kerry are extremely interested in its future. Last year, Senator Bond requested that GAO make a study of the program "so that he could have a full picture about the success of the program and where it needs to go after the year 2000." At a briefing the week before the symposium, GAO "clearly indicated" to Mr. Cooksey that the report would be favorable. There is nonetheless a need for improvement. For example, one question is whether more than 2.5 percent of the R&D budgets of the participating agencies should be set aside for SBIR. Noting that proposals to expand the set-aside might provoke controversy, Mr. Cooksey said that all 18 members of the Senate Committee on Small Business believe in the program and predicted that not a single member of the Senate would oppose renewing it.

DISCUSSANT

James Turner
Minority Counsel
House Committee on Science

Mr. Turner pointed out that he was attending the symposium with his Republican counterpart on the House Committee on Science, Scott Geesey. Stressing that he was not speaking on behalf of any particular member of the committee, Mr. Turner said that, contrary to some perceptions, he does not believe that the House is out to terminate the SBIR program. He said that the House Committee on Science and the Senate Committee on Small Business do differ in their definitions of the SBIR program and the ways in which it should be improved. He expressed the hope that the two Congressional committees would "proceed in good faith to try to work out any differences in a friendly way."

Focus on SBIR Results

Mr. Turner stated that the Government Performance and Results Act (GPRA) provides the current "context of improvement to programs" of the federal government and recalled that an amendment passed with the extension of the Small Business Technology Transfer program made the latter, like most other federal programs, subject to GPRA, obliging it to establish performance goals and to measure its results against those goals. "I think this is the least we can ask of any program," he said, indicating that a similar measure would be proposed in conjunction with the SBIR program's reauthorization.

Program Scale

SBIR is notable for its sheer size. Even though SBIR receives only 2.5 percent of agency R&D budgets, it may be the largest federal grants program that exists. Its annual funding (over $1.1 billion) probably exceeds the total expenditure on the Advanced Technology Program (ATP) since that controversial program started up in 1990, and it likely amounts to twice the overall funding of another program managed by the National Institute of Standards and Technology, the Manufacturing Extension Partnership. Although it may be "small by budget standards, it's not small by appropriations standards or authorization standards," Mr. Turner said. He advised those in attendance to "expect the level of scrutiny that a big program deserves."

The Need for Evaluation

In light of its size, SBIR is, in his view, "a stealth program in some ways." There has been far less formal evaluative work done on the SBIR program than on the much younger ATP, and far less effort has gone into employing expert advice to perfect the former program. Yet, since SBIR's start up in 1983, small business, collaborative research practices, the ratio of federal to private-sector investment in R&D, and technology itself have all changed significantly. Mr. Turner cited a recent study by the Berkeley Roundtable on the International Economy indicating that focusing on technology transfer may now be as important a role for the federal government as funding research.

Potential for Program Improvements

Although conceding that the base of the SBIR program is not likely to change dramatically, he said that attempts to improve it should focus on:

- whether the program has been a success and finding evidence "beyond anecdotes" to measure its success. What share of government contracts did small business hold when the program started, and what is its share now? If there have been gains, how many of the gains can be credited to the SBIR program?

- whether innovation truly is concentrated in firms with 25 or fewer employees. If so, should firms with up to 500 employees that may have subdivisions and joint ventures be considered small businesses? Is there a need for rethinking what type of company should be funded under SBIR?
- whether more extensive federal–private partnering might ensure and speed the commercial success of SBIR award winners. Should applicants be required to submit business plans from the beginning, as they are under ATP? Should agencies make their research capabilities available to their SBIR awardees?
- whether integrating SBIR into the technology transfer programs of its sponsoring federal agencies would help insulate technologies developed by SBIR winners from alleged cream-skimming on the part of foreign interests.
- whether a conscious effort should be made to tie SBIR into the large number of state and local small business economic development programs that have come into existence since 1983.

Conflicting Program Goals

Mr. Turner then addressed what he called "the schizophrenic nature" of the SBIR program: Awards are required to support research that meets the needs of the awarding agencies, yet products developed under the program are expected to be successful in the commercial marketplace. Thus, there may be a conflict between the goals of procurement and innovation. He called for a broader examination of the nature of government agency needs and a keener attempt to identify the end users of products developed with SBIR funding. If the client is going to be the federal government, "maybe what we need to do in terms of partnering and support is to make sure that those small businesses are getting into the supply chain and the programs for the federal government," he said. "Maybe, if there is going to be an increase in funding for small businesses, we need to be looking at the way that they get a bigger share of the pie." Although acknowledging that some agencies may already have undertaken such efforts, he proposed taking a more systematic approach.

Similarly, attention should go to determining how firms whose products are destined for the commercial market can be linked into private-sector supply chains and to ensuring that large corporations take SBIR firms seriously. Recalling hearings that took place before the SBIR program's creation, Mr. Turner noted that the Big Three automakers indicated they "just would not touch" breakthrough ideas under development in small firms out of fear of patent problems: The larger corporations "did not want to listen to an inventor and then somehow be prejudiced if one of their people came up with the same idea." Programs like SBIR can prove valuable by facilitating mutually beneficial contact between the ultimate user of a product and its inventor, thus closing this gap in the innovation system.

Summing up, Mr. Turner said that the House Committee on Science does not consider itself an enemy of the SBIR program. But unlike the Senate Committee on Small Business, which is the official advocate for small business, the Science Committee, as the official advocate for research and competitiveness, has a broad constituency that includes universities and larger companies as well as small business. "We do not think of small business as an island," he said, "but we would like to work with [the small business] community to find ways to create win–win situations for America and for its small businesses as well."

DISCUSSION

Opening the discussion period, Mr. Neal called on Robert Carr of the Science and Technology Policy Program at SRI International, who asked whether there is a significant number of firms that have won multiple SBIR awards and, if so, whether that is positive or negative for the program. In answer, Mr. Turner said that one company has won over 300 awards in the 1990s, that other companies are close to that number, and that a sizable number of companies have won 50 or more awards over the life of the program; many more companies have won a single award or a few awards.

Amendments addressing the case of repeat awards have been offered in the course of the program's last two reauthorizations, indicating that it is viewed by some as a problem, and Mr. Turner said that he would not be surprised to see such amendments in the upcoming reauthorization as well. The phenomenon is evidence of the program's schizophrenic nature, he argued, contending that many award winners are working as "government contractors rather than innovators," and stating, "You have to decide what your real purpose for the program is before you know whether it's a problem or not."

Agreeing with Mr. Turner, Mr. Cooksey said that the Senate Committee on Small Business expects information from GAO on the magnitude of the problem—if in fact a problem exists, or exists for agencies other than the DoD. The committee will also look at other questions, such as one raised three years ago by GAO: multiple payments to the same contractor for the same project. Ms. Forbes cautioned against doing anything that might change the competitive nature of the program, to which she attributed its success.

Mr. Neal noted that the DoD SBIR program has recently instituted a Fast Track pilot program designed to address the problems raised by Mr. Turner and Mr. Carr with the ultimate goal of ensuring that products under SBIR become commercial. The desire to ensure greater success in commercializing products led to the institution of the Fast Track pilot, but it has been in place for only a short time and the results are still under examination. Defense Department officials are working with Congress to determine performance measures for the SBIR program.

Martin Apple of the Council of Scientific Society Presidents expressed concern at "how Congress in particular, and the whole of Washington, D.C., attempts to manipulate the innovation systems of the country without understanding them." Calling the overall innovation system very delicate, he claimed that how such programs as SBIR and ATP fit into the system is not well understood, and he warned that attempting to improve the system without fully understanding it might do more harm than good.

E. Martin Duggan of the Small Business Exporters Association asked whether any data exist that indicate SBIR participants' level of success in international trade. Mr. Cooksey acknowledged that there is a lack of good data on this and said that his committee would bring up the matter with GAO. Addressing the comments of both Dr. Apple and Mr. Duggan, Mr. Turner said that, with reauthorization of the SBIR program, he does "not think anyone believes that the program is not going to be renewed, and [has] not heard anyone who has said that the program can not be improved." Most valuable at this time would be suggestions as to which aspects of the program should be studied to help Congress, which he expected to "tinker at the edges [of SBIR] more than turn the program on its side." He encouraged those in attendance to pass along their ideas to him and his colleagues in the ensuing weeks. Mr. Tibbetts, referring to his review of 50 awards, said that 42 of the firms exported and that exports accounted for 34 percent of the $9.1 billion in direct and indirect sales attributed to the SBIR program.

James Woo of the Small Business Technology Coalition praised Mr. Turner as a friend of the SBIR program for highlighting what he also sees as its "schizophrenic nature." That all agencies attempt to run the program using the same metrics constitutes a problem in light of the agencies' differing missions. Dr. Woo called putting an emphasis on commercialization at the DoD or NASA "just the wrong metrics," whereas fostering economic development is an important goal for NSF or the National Institutes of Health. He urged participants to reflect on how the program's structure might be changed so that each agency could employ appropriate metrics in measuring its success.

Panel II

Research Perspectives on the SBIR

INTRODUCTION

Zoltan Acs
University of Maryland

Dr. Acs opened the session by noting the recent publication by the Small Business Administration (SBA), *State of Small Business*, that is an assessment and overview of federal programs designed to stimulate the growth and development of "new-technology-based" firms. Programs examined in the publication included the SBIR program, the Small Business Technology Transfer program, several financing programs managed by SBA, along with the Commerce Department's Advanced Technology Program (ATP) and Manufacturing Extension Partnership. There is, however, a dearth of academic studies examining the impact of programs under which the federal government and the private sector have formed partnerships to develop and deploy new technologies. Most of these programs, he noted, built on the experience of the states and have been instituted since 1980. In this context, Dr. Acs pointed to the value of work to be presented by the panel's main speaker, Josh Lerner of the Harvard Business School, on the role of the SBIR program.

THE GOVERNMENT AS VENTURE CAPITALIST: AN ANALYSIS OF THE SBIR

Joshua Lerner
Harvard Business School

After spending most of the past decade studying issues associated with traditional venture-capital organizations—such as how they finance high-technology companies and what the impacts on innovation are—Dr. Lerner has turned his attention to the interaction of public policy and the entrepreneurial sector. This is the area of concern that occupied him during the early years of his career, spent working on technology policy issues in Washington, and a recent surge of interest both in the United States and abroad has drawn him back to it.

Public Support for High-Technology Start-Ups

Dr. Lerner began the presentation of his work by describing the size of public-sector support for entrepreneurial firms relative to that provided by private venture capitalists. Although public programs no longer dominate the field the way they did in the 1960s, when their combined size was three times that of the private venture funds, their resources still amount to a "significant fraction" of those of an expanded independent venture sector. In addition, there is much evidence that government programs such as the Small Business Investment Corporation (SBIC) program retain considerable importance, whether in the list of firms that have received money from SBICs or SBIR when they were still privately held and relatively unknown or in the experience that many leaders of today's U.S. venture industry received under SBIC several decades ago. The growth of venture-capital sectors in countries such as Israel and Taiwan, where the role of government policy and programs in encouraging entrepreneurial activity has been significant, reinforces this picture.

Dr. Lerner concurred in the view that SBIR has received little attention or scrutiny for a program of its size. Qualifying his own study as preliminary in that it concerns a limited set of questions, he said it attempts a systematic look at the growth and evolution over a decade of both the early SBIR awardees and a matched set of firms that have many, although not all, of the same characteristics. In particular, he was interested in ascertaining whether any specific subset of SBIR firms derived exceptional benefit from the awards.

Study Summary: An Effective Program

Dr. Lerner said that his study suggests that the awards generally make a difference in that the award-winning companies are "associated with greater growth." The greatest growth, however, is associated with awards made to companies in regions of the country with significant entrepreneurial activity in high-technology industries. He also found that the firms that won the fewest awards displayed growth equal to or greater than that displayed by firms that won multiple awards. The picture that emerges, he argued, is of a program that is working effectively and seems to be playing an important role in stimulating young firms; much of the evidence is consistent with the argument that SBIR awards are important in providing young firms with certification, a "stamp of approval" they can then use in obtaining access to other resources. At the same time, Dr. Lerner allowed, it may well be possible that "some distortions in the awards process" exist and that they should be further assessed.

Rationale for Subsidies

Before taking up the data, Dr. Lerner proposed examining the rationale, from a policy perspective, for offering awards and subsidies to small, high-technology companies. He cited two arguments that economists make in favor of them:

- **Spillovers or externalities:** Many of the benefits of R&D programs do not accrue uniquely to those undertaking the research but are captured also by consumers and by firms developing complementary or competing products. This problem may be particularly acute for small firms with relatively limited resources, which may not have sufficient incentive to undertake research and for which subsidy programs may thus play a significant role.
- **Information problems:** Investors are frequently hampered by a lack of technical expertise in assessing the activities of small, high-technology companies, for which reason viable technologies may not attract the funding they merit. Though theoretically designed to deal with this challenge, venture capital represents a "very narrow and targeted financing mechanism." To illustrate, Dr. Lerner cited figures for 1996, during which roughly 1 million firms began operation in the United States: Although this was a record year for venture-capital activity, only about 600 firms received venture financing for the first time. "Of these million companies, many were mom-and-pop convenience stores, solo consulting practices, and so forth, which did not need the financing," he acknowledged. "But even after you throw them out, there is still a very substantial number of companies relative to the very small number that received venture financing." Another manifestation of the venture sector's lack of expertise in evaluating information, according to Dr. Lerner, is a tendency to follow investment fashions that has led it to focus narrowly on a succession of "hot" sectors, such as Internet service providers, search engines, and Internet chat channels. This investor behavior, which may leave interesting and potentially viable companies in less fashionable areas out

in the cold, can to some extent be offset through programs like SBIR that depend on knowledgeable people in many areas of technology to evaluate company proposals.

Striking a cautionary note, Dr. Lerner pointed to economic literature that highlights "real dangers" associated with subsidy programs. Work on "political capture" argues that particular interest groups can succeed in steering benefits to themselves and that programs can be used in ways that may not be to the best benefit of those taking part.[1] Some specific critiques of SBIR have argued that well-connected firms may end up receiving a disproportionate share of awards or that awards may go to companies that are likely to be successful without them, which in turn might distort perceptions of the program's effectiveness. In his own analysis, Dr. Lerner tried to isolate both potential benefits and possible distortions, and to discern some distinct patterns in the relative success of awardees.

Challenges in Assessing Public Technology Programs

Dr. Lerner stated that the two most common approaches to assessing public technology programs—surveys in which the companies involved are asked what the direct benefits of participation in a particular program have been, and academic studies of research funding that take place at the level of the general economy—have not necessarily provided answers that reliably demonstrate whether a program is reaching its intended goals.

Surveys: One problem with the former approach, polling individual award recipients directly, is that a program's advocates may tend to attribute all their success to the program. Conversely, fears of being perceived "to have done too well through the courtesy of public money" might incline an entrepreneur to play down the role of an award that might in fact have been critical; this has been observed in connection with some biomedical inventions, Dr. Lerner noted. Finally, in many instances the award is so closely related to the recipient firm's main activity that ascertaining what can actually be credited to the award is extremely difficult.

Similar problems have been associated with academic assessments, which have tended to focus on the question of "crowding out": whether public funding of research discourages the private sector from funding research on its own. Even though attempts have been made to apply empirical techniques to study the impact of specific programs such as SEMATECH or the SBIR program—so that the question becomes whether a firm's participation in a particular program causes a

---

[1] See for example: Cohen, L. R., and R. G. Noll. 1991. *The Technology Pork Barrel.* Washington, D.C.: The Brookings Institution.

reduction in its own internal R&D expenditure—such techniques may not capture important dynamics of firm behavior.[2] The efforts of small entrepreneurial companies, Dr. Lerner reminded the audience, often revolve around a key researcher, so that adding staff is not going to cause progress to accelerate significantly. Therefore, he argued, "it may be a very rational and very sensible response for firms in the short run to take the money they receive from a program like SBIR and reduce their own spending" in order to push back the moment at which they will run out of funds.

Long-Run Impacts of SBIR

Although both methods of analysis provide some clues to the effectiveness of programs, according to Dr. Lerner, neither addresses the "really critical question" of what happens to the firms and their technologies in the long run. He attempts to elucidate this issue in his own analysis by examining the evolution and success of the firms that received Phase II awards in the first three SBIR program cycles. Relying on data from the U.S. General Accounting Office for an initial survey of these awards, Lerner charted the winning firms' progress over the subsequent decade. He then compared 500 SBIR-winning firms with 900 other firms selected to match them as closely as possible.

Referring to Table V of his paper (see Appendix), he pointed to a comparison of employment growth shown by the SBIR awardees and the matching firms over a ten-year period. Although the two sets of firms were roughly at the same size at the end of 1985, a decade later the SBIR awardees had grown by an average of 26 jobs compared with 5 or 6 jobs for the matching firms. On this basis, he saw a significant differential both in job creation and in sales growth between the awardees and the nonawardees. This differential was magnified in geographical areas where the level of venture-capital activity was high in the 1980s. In regions of the country boasting much high-tech entrepreneurial activity, employment at SBIR-winning firms grew by 50 jobs as compared with 3 jobs at non-SBIR firms; the differences discerned in other regions were not statistically significant. From these data, Dr. Lerner posited a "fundamental pattern" according to which the awardees experience far more growth and the growth is concentrated in high-tech regions and industries; scrutiny of multiple awardees revealed no additional growth to be associated with the additional awards.

Besides looking at how well individual firms did under SBIR, it would be desirable to examine the impact of the program on society through its effect on the rate of innovation, on the growth of knowledge, and on spillovers to other firms; doing so would be very difficult, however. Also useful would be compar-

---

[2] D. A. Irwin and P. J. Klenow, 1994, *High-Tech Subsidies: Estimating the Effects of SEMATECH*. NBER Working Paper No. 4974. Washington, D.C.: National Bureau of Economic Research.

ing crucial aspects of the SBIR program with similar programs, such as ATP. In Dr. Lerner's judgment, the SBIR program, substantial as it has been, has lagged ATP in evaluating the research function. He encouraged members of the SBIR community to reflect on and, perhaps, emulate ATP's efforts in "bringing in people from policy as well as economic perspectives to try to encourage studies—many of which have been thought-provoking and, in some cases, quite critical." The result, he said, has been "a very healthy dialogue."

DISCUSSANT

*Kenneth Flamm**
The Brookings Institution

Congratulating Dr. Lerner on the quality of his paper, Dr. Flamm began by listing four topics he wished to discuss in his presentation:

- the fundamental, underlying rationale an economist might offer for inventing such a program as SBIR;
- Dr. Lerner's empirical work;
- Dr. Flamm's personal experience with the program during his service at the DoD, which might suggest further questions for Dr. Lerner's study; and
- the issue of social valuation of the SBIR program, in particular the issue of crowding out in R&D, with special reference to the study of SEMATECH mentioned by Dr. Lerner, as Dr. Flamm believes that both the study itself and the manner in which it is often interpreted are seriously flawed.

The Rationale for SBIR

In addressing his first point—which he recast as "Why SBIR?"—Dr. Flamm recapitulated two answers offered by Dr. Lerner:

- that the social return of R&D projects has often been shown to be greater than their private return, in part for reasons connected with the extent to which private investors can appropriate the benefits of innovative activities; and
- that information asymmetries in financial markets create a role for government as a certifier of worthwhile projects.

---

*At the time of this symposium, Dr. Flamm was a Senior Fellow at Brookings. He now holds the Dean Rusk Chair in International Affairs at the LBJ School of Public Affairs at the University of Texas at Austin.

Social Returns: Commenting on the former, Dr. Flamm pointed out that if the fundamental and underlying rationale for government involvement in or subsidy of technology programs is seen to be the large potential gap between social return and private return and if the market can be counted on to provide information on private return only, not on social return, then there is a "natural argument" to be made that government agencies, whose objectives are social by definition, might have expertise to offer in identifying areas of high social return and in steering projects toward those areas.

In fact, a program like SBIR that targets particular projects has some advantage over some other proposals for capturing social return, Dr. Flamm argued. One such proposal, offered by Scott Wallsten of Stanford University, is using R&D tax credits rather than SBIR to subsidize investment in innovations for which social return is expected to exceed private return. The problem with this idea is that a project with a small private return will have a small private return even after a tax credit; in contrast, a project offering a private return but no social return will have a larger private return after the tax credit. So projects for which low private but high social returns can be expected "are not going to be thrust over the [investment] hurdle rate" by the prospect of a tax credit, which instead would encourage "excess" investment in projects that promise reasonably high private returns but little or no social return. Dr. Flamm thus characterized the tax credit as a "blunt instrument" when it comes to identifying areas with high social returns and trying to get projects in those areas funded; in contrast, he argued that, at least in theory, [the SBIR program] has some potential for funding some projects with higher social returns.

Information Asymmetries: Praising Dr. Lerner's discussion of the question of information asymmetries in financial markets, he connected the notion that the role of government as certifier of a project's value might be particularly important for small firms to another phenomenon that he judged to be worth exploring: While large firms should be able to fend off litigation attacks on their intellectual property, small firms lack the basic resources to defend themselves in court.

Suggesting a rationale for the SBIR program not mentioned by Dr. Lerner but often invoked by those who visited him at his former office at the Pentagon, Dr. Flamm posited the existence of an imperfection in capital markets stemming from the limitation of liability under the bankruptcy laws. The only alternative to venture capital for a small company seeking to raise financing is to seek a bank loan; but lenders require collateral, "and if your collateral is your energy and ideas, then—because the institution of slavery is illegal—there is no collateral if you go bankrupt."

Empirical Issues

Addressing the second of his four original points regarding Dr. Lerner's findings, Dr. Flamm commented that Dr. Lerner's paper points to a "thought-provoking" correlation, at least on a regional basis, between receipt of SBIR funding and commercial success. Warning that correlations do not necessarily imply causation, Dr. Flamm pointed to three possible explanations for the correlation, although he stressed that none is definitively established on the basis of Dr. Lerner's work:

- Government picks winners. "Contrary to popular belief, government does an excellent job of picking winners; that is, the government bureaucrats who are picking these projects are astute enough to pick the firms that are going to succeed anyway."
- Government causes winners. There are three different possible mechanisms: Government certification causes success in dealing with asymmetries in the financial market; giving the actual dollars causes success, although Dr. Lerner appears to have debunked that by showing that companies receiving more dollars do not seem to have a higher success rate than companies receiving fewer dollars; and further contracts from the government cause success, whether those are procurement contracts or R&D contracts.
- Winners pick the government. There are certain characteristics that lead firms to apply to the SBIR program—among them, perhaps, firms that know they are going to be qualified to win this type of grant end up applying and winning.

Experience at DoD

Many Programs: Turning to a consideration of what he learned through direct experience at the DoD, his third point, Dr. Flamm stated that SBIR is not a single program but, in fact, "a different program in every government agency." Because the methods used to pick the grantees, although subject to general guidelines, may vary from agency to agency or even from one period to another within a single agency, it is important to specify the source of the funding when assessing the program.

An R&D Tax: Second, he recalled that he and his colleagues in the Office of the Secretary of Defense (OSD) saw the SBIR program as a tax on agency R&D budgets. They held as their "deepest, darkest suspicion" the fear that the program was "wired": "Somewhere out there, a director of an R&D lab at the Army or the Navy was coming up with topics that his ex-employees or buddies were going to pursue using SBIR grants; and the topic would be written in such a way that Dr. X, 'retired from the Naval Surface Tank Division' or whatever, was going to get the grant." This anxiety induced OSD officials to seek ways of increasing the visibility and transparency of the SBIR program. Still, Dr. Flamm conjectured that some of the program's apparent success may actually be resulting in the "incubation of 'Beltway Bandits'"—that firms getting a start through the SBIR program might subsequently stay alive as contract researchers by cultivating close relationships with government agencies. He suggested that it might be possible to determine whether this is indeed happening by looking at the share of revenues of SBIR-winning firms that come from the government, and that this subject would be an "interesting" one to consider.

Social Valuation

As for the last of his four issues, that of social valuation, Dr. Flamm noted Dr. Lerner's even-handedness in referring to a study by Irwin and Klenow that concludes that the main effect of federal funding of the SEMATECH consortium is the reduction by private firms of their spending on R&D.[3] He declared, however: "I would like for the record in a public forum to point out that there are some fatal problems with that study that never get mentioned." He said that there are two specific errors in the study, but he had time to speak about only one of them.

Using a sample drawn from Standard & Poor's Compustat database of semiconductor companies, the authors distinguished between those that belonged to SEMATECH and those that did not. But more than half of the non-SEMATECH firms in the sample were makers not of semiconductor devices but of semiconductor equipment and materials. The companies are classified together with the chipmakers because there is no separate Standard Industrial Classification category for semiconductor equipment and materials companies. According to Dr. Flamm, however, if the latter are included, it skews the sample. Until the study is redone making a "proper parsing of the data," he recommended suspending belief on what the potential conclusions might be.

DISCUSSION

Dr. Acs opened the discussion by recounting, in support of Dr. Lerner's arguments, two findings from a paper he had recently published on local geographic spillover between university research and high-technology innovations: (1) Most innovations by small companies that involve university research take place very close geographically to the university. He cited the distance as 50

[3] Irwin and Klenow, 1994, *op. cit.*

miles, which he defined as "driving distance in Boston in one day." (2) Regions that have high concentrations of large firms or of branch plants of large firms produce fewer innovations. "So there is reasonably strong evidence," he observed, "that small firms, if they dominate an area, will end up being very active in innovation."

Dennis Chamot of the National Research Council posited that an ideal success rate for the SBIR program would be somewhere between 100 percent, which would indicate that selections were overly cautious, and zero, in which case the program would not be yielding any social benefit. He asked what an acceptable rate of success for the program might be. Dr. Lerner suggested that the venture-capital industry provides the best benchmark, because it has had a huge impact on innovation within the industries it has funded, as well as having created a large number of companies that have been able to access the public capital markets. But even though venture-capital firms are extremely selective, funding only one in 100 or 200 of the business plans they consider, they indicate that only one in three to four ventures they fund comes to be regarded as a success. And in the case of even the most established venture-capital groups, the vast majority of returns are generated by a very small number of very highly successful firms, such as Lotus, Netscape, or Genentech. In both the public and the private sectors, "failure is probably going to be the most likely outcome" of investment in small entrepreneurial companies, Dr. Lerner stated. "What one is really looking for is that out of the many failures will come a few that will just be so spectacularly successful that that will undo the many small losses. But that can often be a hard lesson to absorb."

James Woo of the Small Business Technology Coalition pointed out that, as some federal agencies have not funded business per se and as the procurement programs of others—notably the Pentagon—have virtually precluded their going to small business, the SBIR program allows the government to tap into the innovative capabilities of small technology companies in connection with specific agency missions. Spin-offs that result may bring wider economic benefits, said Mr. Woo, dismissing the issue of whether public R&D money crowds out private money as irrelevant.

Lee Mercer of the National Association of Small Business Investment Companies, recalling his involvement with SBIR in its early years, cited two justifications for the program that were current at that time: (1) that small firms could produce innovations more cost-effectively than large firms, and that if the government was going to procure innovation, it ought to do so in the most cost-effective manner in line with the "Reagan mantra at that time: 'We have to do more with less'"; and (2) that, all other things being equal between two competitors for a particular contract, the government should favor the firm that shows the ability to commercialize the technology—not necessarily the potential to succeed as a company.

Panel III

Case Studies

INTRODUCTION

Dorothy Robyn
White House National Economic Council

Speaking as the National Economic Council official whose portfolio includes small business and technology issues, Dr. Robyn noted that the Clinton administration has been very supportive of the SBIR program. She recalled that President Clinton and Vice President Gore campaigned in 1992 in support of legislation doubling SBIR funding, and she said that the administration has consistently backed the program's extension. She nonetheless recognizes that "there is an increasing level of concern, resistance, [and] hostility in some of the agencies toward SBIR." Traditionally, she pointed out, SBIR has been viewed within some executive branch agencies as a tax on the R&D budget; over the past few years, as the SBIR set-aside has grown but R&D budgets have not, some agency officials who had been content to overlook the program's impact on their resources have become increasingly concerned "about the size of the pie that [SBIR] is taking."

Dr. Robyn expressed her own view that this development is an opportunity rather than a problem. DoD has transformed its SBIR program, which had had the reputation of being weak, under the leadership of Dr. Flamm, Jon Baron, and Mr. Neal; some other agencies are moving in the same direction. When Phillip Lader was head of the Small Business Administration and Mary Good was under secretary of Commerce for technology, the Clinton administration started an inter-agency effort aimed at propelling internal reform of SBIR that would "make it more driven by agency missions"—and, consequently, help achieve greater buy-in on the part of the agency officials who run it. This effort has lapsed, but Dr. Robyn saw the current symposium as a chance to review some of the ideas that were the basis of SBIR reform early in the Clinton administration. Cautioning against a "SBIR love-in," she called on the symposium's participants to offer constructive criticism as the only way to ensure progress.

RELATIONAL TECHNOLOGIES

Gary Morgenthaler
General Partner
Morgenthaler Ventures

Dr. Morgenthaler introduced himself as a past recipient of SBIR grants and the founder of a software company spun out of University of California at Berkeley that was formed in 1980, taken public in 1988, and sold in 1990. For the past eight years he has been general partner of a private equity venture-capital firm in California that mainly funds high-technology ventures. With a new $200 million fund now in formation, the firm will manage $500 million in capital. The fund has financed about 150 companies, most of them in the fields of information technology and health care; 34 of the last 62 firms the fund has invested in are today public companies. "We are early-stage investors," Dr. Morgenthaler said, "but we fund high-growth companies and we build companies that target public markets."

Personal Lessons from SBIR

Ingres, a pioneering relational database management systems software company founded by Dr. Morgenthaler nearly two decades ago, received SBIR grants totaling $650,000 in 1984, 1985, and 1986. It grew to 1,300 employees "from a raw start-up" and posted $160 million in revenues before being sold; today, as a division of a larger company, it has revenues in the range of $200 million. The company has generally been viewed as a success, having achieved 100 times return on initial capital and ten times overall capital invested. It has produced three spin-out firms whose combined market capitalization has climbed above $1 billion. Boeing was a partner in implementation, and its AWACS (airborne warning and control system) program used the technology in the Gulf War, evidencing that the technology delivered "lots of social returns as well as economic returns."

Among the lessons his company learned through its involvement with the SBIR program, Dr. Morgenthaler includes the importance of grantsmanship: the art of figuring out which grants will be awarded and which will not. But, he stressed, his experience taught him that "if you delivered on your promises, the

SBIR was fairly benevolent and provided support to the companies that really took the program seriously."

The Role of the Venture Capital Industry

Dr. Morgenthaler noted that the venture-capital sector boasts 600 firms with 2,500 partners in North America and that its market has grown from about $4.5 billion 18 months ago to a projected $9 billion for 1998. Because "angel" investors fund ten times what venture-capital firms fund, the venture-capital sector may in fact be viewed as the tip of the investment iceberg. Still, between 20 and 30 percent of initial public offerings (IPOs) are venture-capital backed, and the U.S. venture-capital industry is "the envy of most countries of the world" because of its success in creating both jobs and wealth. The Silicon Valley model is being emulated not only in Boston, New York City, Seattle, Austin, and northern Virginia, but in the United Kingdom, Sweden, Israel, Taiwan, and Singapore.

Citing a recent study by Coopers & Lybrand, Dr. Morgenthaler noted that venture-backed companies displayed annual sales growth of 38 percent as compared with 3.5 percent for Fortune 500 companies. Similarly, job creation was 33 percent annually in venture-backed industry versus –3.6 percent for Fortune 500 companies; venture capitalists supported firms that had four times as many engineers and scientists in management as the U.S. average. Compared with the national average, venture-backed firms have three times the R&D intensity, four times the R&D dollars per employee, and 2.5 times the R&D dollars per dollar of equity. About 81 percent of the venture-capital industry's dollars go to information technology and to health care, including biomedical devices and biotechnology. In Dr. Morgenthaler's opinion, venture-capital firms are investing in the types of firms and helping to create the types of jobs that will be necessary to U.S. economic competitiveness in the coming century.

Explaining that the venture-capital industry cycles in the way other financial services sectors do, he said that its current position—at 2.5 times the peak of the earlier cycle and seven times what the periodic low was as late as 1991—constitutes a boom. Returns have been exceptional: Venture capitalists have delivered to the market about 1,250 IPOs during the current decade, including 20 percent of all IPOs in 1995 and 31 percent of all IPOs in 1996, even though the venture-capital sector represents only about 10 percent of overall private equity investment in the U.S. economy.

Economic Returns to SBIR

Focusing his comments on the area of economic rather than social returns, Dr. Morgenthaler said that, although the SBIR funds many worthwhile programs, "like all government programs [it] is subject to distortions and even to abuse."

He referred to the "SBIR mills" that have received between 50 and 300 awards, calling the figure "slightly mind-boggling" and saying that, in such cases, "tax-payer value tends to get obscured." He then proposed that closer collaboration between the venture-capital industry and the SBIR might benefit the program, arguing that the venture-capital community's objectives and those of the SBIR program overlap directly.

"My proposal is that good science may be a necessary [criterion] but is not a sufficient criterion for evaluating SBIR grants," Dr. Morgenthaler stated. "Growth companies need investor capital, but growth companies also need management, and if you evaluate your grants solely on the basis of good science, you will find that these are companies that do not build viable commercial entities and take that science to the marketplace." Venture capital, in contrast, is instrumental in building companies: "Venture capitalists recruit senior management, including CEOs, for these companies; they help set market strategy; they syndicate financings; they broker corporate partnerships; and they oversee the daily hard work of corporate governance."

But although venture capital funds development, it funds very little research, even though many of the companies it supports are capable of doing first-class research. SBIR plays the role of enabling small companies, including venture-backed companies, to undertake high-risk research projects. "By placing its grants alongside those grants from the private equity industry," Dr. Morgenthaler stated, SBIR "maximizes the chance that excellent research will be coupled with effective management and adequate financing."

Areas for Improvement in SBIR

He then named several key issues he believes SBIR needs to address:

- Timing. The 18 months that can pass from an initial proposal to a Phase II grant is, in today's context, "an eternity," during which an idea can lose its competitive value as the market moves on.
- Social returns. Although critical, this issue is one about which the venture-capital industry has little to say.
- Peer review. Dr. Morgenthaler suggested that SBIR look to the venture-capital industry in evaluating proposals that are not of overwhelming social value but might still be worth supporting in and after Phase II.

DIGITAL SYSTEM RESOURCES

Rich Carroll
President

The Need for Innovation at DoD

Mr. Carroll began by declaring his support for leaving the SBIR program unchanged, explaining that his firm, which developed new SONAR for the U.S. submarine fleet, has been successful under the program in its current form. He then proposed to speak, from this admittedly "biased" position, about his personal experiences with SBIR in order to contribute one answer to the question, "What is the story behind the story of each [SBIR] success?"

Referring to a copy of the DoD budget, Mr. Carroll pointed out that, although the department's R&D allocation is down 14 percent this year, the Pentagon still has around $36 billion to spend on R&D annually—half of the federal government's total expenditure on R&D and one-quarter of the nation's. In light of this, he argued, no one should be shocked to learn that a DoD contractor like his firm is a recipient of SBIR awards. "It's very important to get innovation into the DoD," he stated. "I think that the DoD needs innovation to protect our nation more than our commercial sector needs innovation to improve the quality of life in our nation."

Commercialization at DoD

The commercialization focus of the SBIR program is one area in which Mr. Carroll's company has experienced some difficulties: It has had submissions turned down in both Phase I and Phase II because it attempts to commercialize only within DoD. According to proposal evaluations received from the department, he said, his company has been deemed ineffective in commercializing its products despite the fact that it has sold products to the DoD for $50 million per year that were developed with the aid of the SBIR program, and despite the fact that 50 percent of the Phase I and Phase II awards it has won have resulted in sales. He attributed this negative judgment to a perception within the program that "commercialization" refers only to sales to the private sector, and he recommended guarding against the spread of this attitude throughout the SBIR program.

Strategies for SBIR Proposals

Mr. Carroll outlined how he goes about submitting a SBIR proposal and attempting to commercialize it within the DoD. First, he holds "innovation focus groups" about every two months in which "innovative people" among the company's 200 employees come together. "We say: 'What can we do to be innovative, [to] come up with new ideas and more affordable ways of addressing the needs of the DoD?'" Each idea is written up just as it would be in an SBIR proposal and handed out to as many people as possible who might be interested in pursuing it as a topic under SBIR. "Innovation," he explained, "really starts at the topic."

Besides working with SBIR officials, Mr. Carroll works with participants in other programs, noting that it is in weapons acquisition that most of the DoD money is spent. He tries to find out who is running such programs as the F-18 jet, the new attack submarine, and other programs that have billions of dollars invested in them; he also tries to find out what their problems are and how technology could help solve the programs. He then attempts to interest the acquisition managers in those technologies because he knows that, "in the long run, they are going to be the customer."

Second, Mr. Carroll bids on the topics that come out, with the firm spending a great deal more money bidding Phase I topics than it receives in Phase I awards. Immediately after winning a Phase I award, he tries to determine what the company should do in Phases I and II to support the acquisition managers. "I pretty much ignore the SBIR topic for now," he stated. "It was one paragraph. It was a good idea, and people are interested in it—but, again, I'm interested in helping the acquisition managers build these weapons systems in the most cost-effective and affordable way."

Thus, when submitting a Phase II proposal, Mr. Carroll already knows that there will be a client for the technology if it is developed. In the event that he has not been able to find anyone who is interested in the technology and willing to put money up to back it, he will not bid in Phase II: In such a case, he does not think he will win and does not want to spend any more time on it. In some cases he has identified a weapons acquisition manager who is interested in buying his technology but he can not get a Phase II award because he fails to meet "some other criteria in the program." Expressing frustration over such experiences, he called for improvement in this aspect of the process. In Phase III, he said, it is important that his company continue to access R&D through the DoD component interested in the technology.

Summarizing, Mr. Carroll underscored three features of DoD's SBIR program as being particularly important. The first is that it gives small businesses an avenue for doing business with the agency. He said that his firm's sale of SONAR technology to the Navy would never have occurred without the SBIR program because no procurement has ever come out for that type of item. The second important feature is that the program provides small firms a customer right from the start. The third is that it gives the firm the data rights necessary to carrying a technology forward and getting the DoD to continue to do business with the firm.

QUANTUM ENERGY TECHNOLOGIES

John Preston
President and CEO

Rather than presenting a case study of one of the SBIR award-winning companies he has started, Mr. Preston said that he was addressing the symposium in the capacity of senior lecturer and co-director of MIT's Entrepreneurship Center. He expressed his desire at the outset to endorse as "absolutely right on target" the results of the paper outlined by Dr. Lerner during the second panel. He added that Dr. Lerner's conclusions have been corroborated by research done at MIT on SBIR awards and on job and wealth creation in the United States.

Why SBIR is Needed

Asking "Why do we need the SBIR program?," Mr. Preston reminded the audience that, although incremental innovation occurs in the nation's large companies, radical innovation does not. A study by MIT's Professor Utterback of 150 radical innovations dating from the past 100 years failed to come up with a single case in which the radical innovation had been pioneered by a market leader. Mr. Preston recalled Western Union's decision in the 1880s to decline Alexander Graham Bell's offer of an exclusive license to the telephone on the grounds that the technology lacked commercial viability, as well as similar cases in which established firms responded to the arrival of a new technology as a threat rather than an opportunity.

"If you look at the 'bang for the buck' that the U.S. government is getting for the dollars that it's spending," Mr. Preston argued, "I would say that basic research and SBIR are way, way at the top." One reason for this, which was established in a study by David Birch that Mr. Preston described as having flaws but that was basically accurate in its conclusions, is that 70 percent of the net new jobs in the United States are being created by only 4 percent of the nation's companies. The overlap between the characteristics of the companies that have created the jobs and those of the companies that the SBIR program tries to finance is extremely high.

SBIR and Time to Market

SBIR has also helped catalyze improvements that have allowed products to get to market faster. To illustrate the importance of this, Mr. Preston turned to the issue of innovation cycles, noting that when a radical innovation is made, it goes through a period of rapid change that is then followed by a period of incremental improvement. The United States is particularly good at the creative end of that spectrum; in industries such as software that are driven by creativity, the United

States does extremely well. However, in industries like consumer electronics that are driven more by incremental improvement, Japan is a very strong competitor.

Over the past ten years, however, U.S. companies have recaptured 20 percent of the global semiconductor industry "totally at the expense of the Japanese." The reason this has occurred, Mr. Preston argued, is that as product half-lives have shortened and the speed at which products must be brought to market has increased, the creative component of a product—in the case of integrated circuits, for example, the microcode used to design the chip—has become more important as a function of time. He reported being told by the outgoing head of technology for Hewlett-Packard, Joel Birnbaum, that getting to the market one month earlier is worth more than all the engineering and development costs of a typical Hewlett-Packard product, whereas six months changes the lifetime profitability of a product by 33 percent.

Addressing the Technology "Funding Gap"

Mr. Preston argued that SBIR, by boosting spending at the base, may increase the probability of technologies making it through the "funding gap" that exists in many U.S. industries, particularly those connected with the physical sciences. While in the United States the government spends on basic research and industry spends on commercialization, very little investment is made on the first economic proof of concept. Acknowledging that many would contend that this funding gap is not real because the free market evens it out, Mr. Preston cited the export of U.S. technology—its loss of inventions made in its own laboratories to foreign competitors—and insisted that other countries are better at commercialization.

One example among many is the hard-disk drive, a product that was invented in this country; Singapore now accounts for 70 percent of global disk drive production. Commenting on the national technology plan of Singapore, on whose National Science and Technology Board he has served, Mr. Preston said that Singapore and other countries are "picking winners very openly" and investing very heavily in the funding gap. Meanwhile, the emphasis of U.S. venture funds has, over the past ten years, shifted to the later stages of investment. Calling the SBIR program and the Commerce Department's Advanced Technology Program early-stage investors, he praised them for "attacking a need at the earliest, where it's really the deepest in the trough."

Suggestions for Changes to SBIR

One potential improvement in the SBIR program suggested by Mr. Preston is increasing the number of Phase II awards at the expense of Phase I awards. In attempting to finance an innovative technology, a company can either inject a small amount of money over a long period of time or be more aggressive and investing that money in a shorter period of time. He supported his proposal to consider shifting the relative weight of Phase I and Phase II investments with the contention that companies which do very few things very well and make a few aggressive investments outperform those which do many things in a mediocre way.

Finally, Mr. Preston cautioned against distributing SBIR awards geographically for political reasons; he said that the awards should go to areas where "clusters of excellence," as defined by the Harvard Business School's Michael Porter, are going to support the building of innovative companies. Taking an example from Dr. Porter's work, he pointed to the fact that Holland controls 70 percent of the cut flowers sold in Europe and ships flowers to Disney World in Florida on a daily basis. To grow flowers, Mr. Preston observed, "you need sun, rain, and land—and Holland has only one out of three, rain. So why are the Dutch shipping flowers to Disney World?" The answer, he argued, is that the country has a cluster of companies, along with expertise which supports them, that enables it to be far more competitive globally than one would expect.

DISCUSSANT

Lance Davis
Department of Defense

Dr. Davis, deputy director of the DoD office that runs a part of the agency's SBIR program, opened his remarks by stating that he would outline some of the challenges inherent in the way SBIR tends to be perceived by participating agencies.

Background on SBIR Program Changes

Because DoD's budget for research, development, testing, and evaluation (RDT&E) is used as the basis for calculating its SBIR budget, in FY 1997 the agency's SBIR budget was $537 million, whereas its Small Business Technology Transfer (STTR) program budget totaled $30 million. Concurring in the widely expressed judgment that the department's SBIR program is much improved, Dr. Davis stated that improvement dates to the tenure of Anita Jones as director of Defense Research and Engineering, when the office decided to improve its interaction with small business by looking at small business as if the latter were a customer. Some current features of the program—among them, the Fast Track pilot, the posting of topics on the World Wide Web, and the opportunity for a potential proposer to speak with the author of a topic prior to submitting a solicitation—had their beginnings in focus groups conducted with representatives of small business in different parts of the country. Although praising Mr. Baron, Mr. Neal, and Mr. Hill for contributing to the program's improvement, Dr. Davis stressed that there are in fact eight separate programs within the DoD's decentralized SBIR framework and praised the program managers at the various services and agencies within the department for making the program work.

DoD Selection Criteria

Dr. Davis pointed to an emphasis within the department's SBIR programs on two selection criteria: (1) whether the technology has dual-use application, which is expected to maximize the possibility of commercialization in the long run; and (2) whether a company has a commercialization plan—or there is at least an indication that the applicant can commercialize the technology, either for the defense or the private-sector market. "It's not that easy for all small businesses to make a living in the defense business, because the market is relatively small [and] the number of platforms we are building today is relatively small," he said, so "there's some bias on our part to look at technologies that we think can also be commercialized in the private sector." If technology is not commercialized in the private sector, it is unlikely to remain available to the department, an eventuality that would call into question the value of developing it in the first place.

Future Challenges

Dr. Davis concluded his remarks by pointing to three problems:

- *Program size.* The SBIR "tax," as it is defined by DoD's comptroller and thus perceived within the department, is 2.5 percent not only of the agency's science and technology activities but of the overall RDT&E budget. When program managers of weapons platforms, whose activities are far removed from research, discover that 2.5 percent of their budgets is going to SBIR, he said, "we have a tremendous amount of push-back from those program managers." Stating a personal view, he cautioned that any campaign to enlarge SBIR significantly would occasion vigorous resistance from the program managers "because they think the tax is becoming too big."
- *Funding mechanism.* Dr. Davis advocated finding another way to fund SBIR, even as he acknowledged the suggestion's limited practicality: If the program were supported by a yearly appropriation, the department itself would cut it. Noting that "constituencies on [Capitol] Hill have discovered that a taxed program is a marvelous stealth program because you do not have to justify it every year, you only have to renew it every four or five or ten years," he recounted that there was a call last year for a "tax" to support manufacturing technology and that there has been discussion of a similar scheme to fund dual-use technology within the department. "You are going to come to the ridiculous situation in which 130 percent of your science and technology budget is fixed for something and

you have no flexibility," he predicted. He added: "One of the things the Department wants to be able to do is to respond to technology challenges, and if we have too many things that box us in, it's going to be a problem."

- *Administrative expenses.* Citing the value to a SBIR awardee of being able to speak with the program manager overseeing its project and even to have the manager visit, Dr. Davis pointed out that, because SBIR funds are not allowed to go to administrative expenses, SBIR program activities must compete for overhead funds with all other research activities. "As a result," he said, "the amount of attention that a project director in a laboratory can spend with a SBIR contract awardee is not as great as it should be." If the SBIR program is to be expanded, it would be worth considering returning to the previous practice of allowing some portion of administrative expense to come from the SBIR budget. Simply increasing the tax would not do much to solve this problem, because "as gigantic as the DOD RDT&E budget is, every single dollar has a constituency; so if you take it from one place, it comes from another place, [and] it's just going to be a constant battle."

DISCUSSANT

Richard Russell
House Committee on Science

First Encounters with SBIR

Mr. Russell suggested that recounting how he first came into contact with SBIR might illustrate how congressional staff can come to perceive a program. "Whenever we are trying to fund a program through an authorization, we obviously have to calculate how much money is actually required to fund whatever R&D project we are planning on doing," he said. "I was merrily jotting down some numbers for some legislation when someone informed me that I had to put aside a little extra. This additional program that I knew nothing about, SBIR, actually would strip out part of the money that was going to what at the time was a very valuable R&D program."

Although this story is not meant to reflect on the value of either the SBIR program or the projects it funds, it does illustrate that SBIR is viewed as a tax by people who would otherwise get money that goes into it. This means that SBIR is competing with R&D in all the agencies that contribute to the program. Mr. Russell advised that it is important to keep both of these characteristics of SBIR in mind when discussing either how the program is to be judged or whether it makes sense to increase the rate at which it is funded from general R&D allotments.

Measuring SBIR Results

Also to be kept in mind is that part of the difficulty in assessing SBIR or any other R&D program—and a difficulty with which the Technology Subcommittee has been struggling a great deal recently—is measuring results. Over the past 18 months, as the Government Performance and Results Act has taken hold, every R&D agency has told the panel that it can not possibly conform with the requirements of the act as written because of R&D's long-term, speculative nature.[1] The committee, in discussion with the agencies, has found ways to try to measure what success really is in terms of R&D, although it is not yet certain whether they will work. Although SBIR is no different from any other R&D program in this regard, it is different in that it has a dual function, which makes measuring its success somewhat more complicated: SBIR looks at commercialization as an end while at the same time trying to provide benefit to the agencies that fund it by contributing to their own R&D missions. "Oftentimes, if you are talking about strict commercialization as a measure of success," he observed, "the individual agency that is engaged in funding that SBIR grant may not view that end success as being particularly relevant to its individual needs and its individual mission."

Noting that the process of reauthorizing the SBIR program is in the initial planning stages, Mr. Russell told the audience that hearings on the subject are likely to be held toward the end of the present congressional session as a "warm up" for the reauthorization, which is to take place in the next Congress. Offering an indication of what may happen, he recalled that the STTR program was reauthorized last year with minor changes and with no increase in the set-aside. "SBIR is a much bigger program," he noted, "and will probably have a lot more interest associated with it just because of the dollar value. But at a billion dollars, that's a significant chunk of R&D funding, and coming from the committee that's charged with overseeing at least all the civilian R&D funding for the federal government, it's a responsibility that I know all our members are going to take very, very seriously. The R&D pot, if you look at both defense and civilian, is just not growing that quickly, and so we have to make sure that we are spending the R&D money as well as possible."

DISCUSSION

Mr. Baron of the DoD asked the panelists to elaborate on the possible mechanics of increased cooperation between the venture-capital community and the SBIR program. Mr. Preston responded that the peer-review system SBIR that the uses in evaluating technology may be superior to the evaluation practices of

[1] Committee on Science, Engineering, and Public Policy, 1995, *Evaluating Federal Research Programs: Research and the Government Performance and Results Act.* Washington, D.C.: National Academy Press.

venture capitalists who tend to look at more than a half-dozen business plans each week. On the other hand, he asked, "What venture capitalist would ever, even in their wildest dreams, conceive of investing in a company without interviewing the CEO and the management team and looking at what their track record is, at what the probability of success of those people is in building the business?" Synergy is thus possible between the technological expertise of SBIR program managers and the strengths of venture capitalists, who "can be really good at assessing people and understanding their probability of success—whereas the DoD, I think, is by definition almost unable to make those assessments."

Reacting to the same question, Dr. Morgenthaler pointed out that certain R&D investments that offer social return, such as those in defense and pollution control, are not typically the targets of venture-capital investment. He suggested that greater collaboration might be possible in areas such as information technology and health care and that this collaboration could take several forms. The "simplest and most straightforward" would be in evaluating whether Phase II grants should be awarded based on some inquiry into the potential for commercial success of the awardee company. He agreed with Mr. Preston that it is important to establish whether the managers involved have a track record of developing successful commercial enterprises and whether people involved with the company as directors or investors have experience in building successful companies. He proposed these concerns, which are of importance to venture capitalists, as criteria upon which SBIR awards would in part be judged.

Mr. Carroll commented that if he can not find either a partner outside the SBIR program or a potential customer for the technology to join his company in investing in Phase II of a SBIR award, his company generally does not make a Phase II bid because it views the prospects for commercialization as too uncertain.

Dr. Wessner asked whether different criteria for success should be used in the different agencies that participate in the SBIR program, what an appropriate rate of success for the overall program would be, and whether the success rate should differ according to the agency.

Mr. Preston observed that one criterion for success for the DoD is whether an award will allow purchase of systems at a lower price than if the systems had been custom-made. In general, however, the metrics for success are just now being developed for "SBIR-type" programs, with Dr. Lerner's paper being the first of any significance in analyzing the long-term economic benefits of SBIR. A second problem in answering the question, he added, is that "the economic benefit is longer than the election cycle for people in power in Washington." Looking at the long-term health of the economy would persuade the observer that SBIR offers a very productive investment, but comparing its productivity with that of other investments that the government might make is difficult. Because a key aspect of the program is that it is investing in technologies before industry is fully willing to jump in and after basic research funds would typically run out, Mr. Preston suggested that an indication of success might be whether the United

States is capturing more benefit from the technologies it develops internally and thus reducing its export of technologies.

Tom Goldberg with GHO asked whether the panelists were suggesting that because investors can make money in electronics and biotechnology, the government should confine its support to environmental and defense technologies. He commented that the United States has invested heavily in environmental technologies but, because we have no mechanism to certify their efficacy, we export those technologies to Europe and Asia, only to reimport them. Wondering whether that is "altogether bad," he asked whether U.S. investors should be placing money with those clients abroad that buy the technologies.

Dr. Morgenthaler responded by stressing Dr. Lerner's point that the success of the SBIR program in promoting job and wealth creation has been "collocated with the centers of entrepreneurial activity" and by stating that a critical mass of entrepreneurial excellence and managerial talent is required for that success. "I view venture capital as a catalytic agent," he said, "and I think you should view the SBIR program in the same way: as a catalytic agent." He encouraged managing programs that are focused on social return, such as those in defense technology or pollution control, under criteria different from those applied to programs that are more likely to draw private investment. For programs focused on economic returns, "by leveraging off of the dynamics of the private equity industry, the greatest social good will be achieved."

Mr. Preston praised the question as recognizing that the funding gap does not apply to all industries but is more severe in the energy and environmental fields than in the information or biotechnology sectors where venture capitalists are investing very heavily. Because the SBIR distributed its funds generally, in all areas of interest to the government, it tends to capture more of the emerging industries rather than just reinforcing the industries that experience very little funding gap.

David Speser of Foresight Science & Technology, Inc., asked Dr. Morgenthaler whether dual-use technologies, or technologies that are adaptable to several areas, might not be more appealing to venture capital because they enlarge beyond the government the sphere of clients for a product. Dr. Morgenthaler, answering in the affirmative, observed that the government is increasingly focused on dual use and that its procurement requirements are increasingly focused on commercial off-the-shelf products. Small companies are not ignoring the government marketplace but can not really restrict themselves to it for a variety of reasons.

Luncheon Address

Congressman Tom Davis (R-Va.)

Rep. Davis began by recapitulating the changes in the composition of Congress that have taken place over the past few years, observing that neither the Democratic nor the Republican party has won the entire confidence of the American people. However, the agreement on balancing the federal budget demonstrates that the parties have started to work together. He noted that many in the audience now know roughly where their agencies will stand in the budgetary process over the coming five years, in contrast to the recent past when changes every two years on Capitol Hill or in the White House kept uncertainty high. Now that Congress has shifted and the American people have adjusted, government is starting to change, he said, pointing to the bipartisan Klinger–Cohen procurement bill passed two years ago and to the fact that "the administration and Congress are starting to work in partnership."

THE GLOBAL ECONOMY AND THE INFORMATION AGE

Referring to a statement by the historian Arthur Schlesinger, Jr., that the most important event of the current century has been the invention of the microchip, and after listing the changes that the development of information technology has brought to business practices over the past two decades, Rep. Davis stressed that the economy of the Information Age can be understood only in global terms. The United States feels the impact both of what happens in other nations' domestic economies and of competition in the international marketplace—competition that is particularly significant in the area of technology. It is because of its role in "prepositioning American goods and services and technologies so that the United States will be competitive in a worldwide economic environment," he said, that the SBIR program is so important.

Endorsing the thesis put forward in *The Third Wave* by Alvin and Heidi Toffler—that the major advances in civilization occur when radical, if not revolutionary, technological changes transform society—Rep. Davis concurred in their judgment that the world is now on the verge of an information revolution. "'The industries that have moved into the center of the economy in the last 40 years have as their business the production and distribution of knowledge and information rather than the production and distribution of things,'" he said, quoting Peter Drucker. The new society, a niche society rather than a mass society, will be characterized by specialization and market segmentation. Rep. Davis views congressional actions making "radical changes in the way we do business" in the fields of immigration, taxes, telecommunications, and federal procurement as in line with the arrival of the Information Age.

Particularly with the globalization of the economy, enlightened leaders of both political parties recognize that government must change many things it has done in the past, something successful companies in business and industry are also recognizing. With new inventions appearing almost daily and new research constantly making old products obsolete, policymakers interested in welcoming the future must be flexible; rigid regulation and excessive taxation are "the worst way to encourage the growth of dynamic, forward-looking industries." In this changing environment, in which society is constantly "going back to the source" and encouraging innovation, programs such as SBIR that add a small amount of federal dollars to private-sector resources make a great deal of sense.

THE IMPORTANCE OF SBIR

Currently, more than 6,000 small businesses are involved in the SBIR and the Small Business Technology Transfer (STTR) programs today, and 750 to 1,000 new companies become involved each year. More technically trained employees work in the small business sector than in nonprofit research firms and universities combined. Pointing to the diversity of emphasis displayed by SBIR and STTR, Rep. Davis praised the programs for sponsoring projects in "every conceivable arena of scientific and technical investigation offering truly leading-edge technologies." Small businesses in general represent an extraordinary pool of competence and talent; SBIR companies have used the opportunities provided by the program to identify and develop essential, innovative products not only to their own benefit but to that of the nation.

THE GEOGRAPHIC DISTRIBUTION OF AWARDS

Rejecting the notion that the distribution of SBIR awards should be influenced by state quotas, Rep. Davis stressed the importance of building a critical mass of companies and pointed to areas of concentrated technological activity such as Silicon Valley, Rt. 128, Seattle, Austin, North Carolina's Research Triangle, and northern Virginia. This last area, where his own district lies, today employs more people in the high-technology sector than does Silicon Valley. Such concentration supports not only a salary structure that draws talent but the "constant training and retraining" needed for success; spreading resources around the country would sacrifice many of the efficiencies characteristic of those areas where a critical mass exists. Areas that "have not woken up to the Third Wave and what is needed to start bringing these industries and building that critical mass"—and which, he pointed out, are thus not benefiting from the SBIR program—"get their fair share of the federal budget in many other subsidy areas."

The success of the SBIR program derives from the fact that it does not involve the government picking winners and losers artificially, Rep. Davis said. The program is, instead, a competition of ideas, and it is directed at the small business sector because that is where so much innovation comes from. He warned against complacency; dwindling R&D resources would mean greater competition for whatever research funding is available. And with the current wave of mergers resulting in larger and larger companies, it is vital that the government maintain a pipeline to the innovations that emanate from small businesses. The health of the nation's industrialized economy is fundamentally grounded in successfully converting basic research from the laboratory into technological advancements in the marketplace. The SBIR program affords the government and, eventually, the private sector access to talent and leading-edge innovation for which they otherwise might not receive development funding because benefits spread to the economy through commercialization.

Rep. Davis praised the "fly-before-you-buy" approach inherent in the three-phase structure of the SBIR and STTR programs, which provides small businesses adequate funds for an advanced technology demonstration of their concepts before commitment is made to the technology. In addition, opportunities for follow-on procurements offered through the programs ensure small businesses' enthusiastic participation. Because they are unable to compete with larger companies that sell established products, small firms may die without innovation. The U.S. economy cannot stand still in a world that constantly demands something better, more advanced, and less expensive. It must draw on small businesses to help drive and disseminate innovation.

THE FOCUS ON COMMERCIALIZATION

Referring to the concern that some firms, after winning awards in Phase I and Phase II, do not proceed to Phase III, Rep. Davis said that he does not consider SBIR's success rate a "major problem," even if it merits continued discussion. In fact, he called the SBIR and STTR programs "most successful" in focusing on commercialization, which is what helps the nation over the long term, and he pointed to a "growing consensus" in Congress that they should be kept in place. He called the 2.5 percent of R&D budgets now set aside for SBIR "adequate for the task," expressing his opinion that the size of the total federal R&D expenditure is a more significant figure. Federal R&D spending is currently not as high as it should be if the United States is to maintain its dominance of the world economy in the next century. "The government is going to have to continue to take some leadership role," he said, "but, most important, to incentivize the private sector and give them the tools they need to move ahead." Unlike the federal laboratories, for which commercialization is far from a primary goal, the SBIR program must continue to focus on commercial success, whether in the context of selling to the government or the commercial sector. Stressing the program's value, he urged members of the audience to continue communicating and working with legislators to refine and improve it in the future.

Panel IV

Program Challenges—Operational Views

INTRODUCTION

Dan Hill
Small Business Administration

SBIR's Origins

Mr. Hill opened the session by recalling an exchange that took place at a congressional hearing in 1982, the year the SBIR program was created, between then-Sen. Warren Rudman (R-N.H.) and an official of the National Science Foundation (NSF). The NSF official was complaining about the proposed SBIR as a new set-aside program when Sen. Rudman asked him how many of the agency's R&D awards had gone to small business. The answer was that, in the history of NSF support for R&D, not a single award had been made to a small business entity. Mr. Hill then quoted the Senator's reply: "'Is not it true we are not really creating a new set-aside, we are just destroying the old one?'" Discussions of the "quote-unquote 'tax'" that SBIR funding is said to constitute, Mr. Hill argued, often lose sight of the reason for which SBIR was created: to bring cost-effective, highly innovative small businesses into federal R&D programs.

Second, while backing the call for further examination of the program, Mr. Hill asked that future studies compare its effectiveness to that achieved through the expenditure of the remaining 97.5 percent of federal R&D funds. He objected to what he sees as a pervasive tendency to set a standard of success for SBIR that is "arbitrarily higher" than standards applied to other programs. A comparison such as that he was advocating would yield results favorable to SBIR, he predicted, claiming that small businesses, despite their meager share of federal R&D funding, outperform large companies and universities in obtaining patents.

Concurring with Rep. Davis's judgment that times have changed, Mr. Hill recommended that the coming debate on SBIR's reauthorization not be dominated by the question of how R&D funding is to be split up among the various research institutions. Instead, it should focus on enlarging the R&D funding pie itself, while at the same time affirming the principle that investing in the nation's R&D needs is a good way to spur economic development. The Small Business Administration (SBA) has taken a position in favor of forming new partnerships and "new friendships" with industry and universities. Acknowledging that not all universities support the SBIR program, he observed that a sea change under way in the academic world is being viewed by SBA as an encouraging sign.

Issues in SBIR's Reauthorization

While stressing that SBA believes the SBIR program can be improved, he cautioned against the emphasis on legislative remedies that had been mentioned at the symposium to that point, saying that many of the issues raised can be addressed administratively under existing legislation, either agency by agency or across the board. He expressed his eagerness, as SBA Assistant Administrator for Technology, to hear the suggestions of meeting participants, and he promised to take them up with program managers at the participating agencies. He then concluded by presenting a list of the programmatic and operational issues he believes to be facing the program in the run-up to reauthorization:

- Commercialization: What value should be assigned by the participating agencies to a firm's ability to commercialize? At what point in the process should commercialization potential be considered a factor in awarding a SBIR project? What is commercialization, and does the definition vary industry to industry or agency to agency?
- Cost-sharing: Should SBIR agencies be allowed to decide awards based on the amount of cost-sharing provided by an applicant? Does this really advance the purpose of SBIR?
- Multiple award winners: According to Mr. Hill, authority currently existing at agencies is sufficient to limit those multiple award winners that are not contributing to mission needs or are not active in commercialization. He urged an examination of whether funding from the remaining 97.5 percent of the federal R&D budget tends to become concentrated among a limited number of large corporations and universities and whether a higher standard is being set for small business.

- Time delays: Does too much time pass between the end of a solicitation period and the issuance of a Phase I award and between Phase I and Phase II? How long should agencies take, and is a congressional mandate needed to set that limit?
- Geographic distribution: Is the answer to uneven geographic distribution of awards to set up a quota system, or are there other remedies such as outreach?
- Evaluative criteria: Agreeing that evaluative criteria for both SBIR and the Small Business Technology Transfer program must be improved, he noted that SBA had agreed with the recommendation of the House Committee on Science to make the programs subject to the Government Performance and Results Act (GPRA). In an upcoming report, SBA will provide both the Science and Small Business committees with language that has been instituted by the agencies that participate in the SBIR program to show that they are within GPRA. He called in addition for suggestions from the small business community and Congress on what might be done to evaluate the SBIR program other than placing it under GPRA.
- Three-phase approach: Does this approach make sense and should it be continued? Although he predicted that raising this question would cause "consternation" on the part of many in attendance, Mr. Hill reported that several agencies had approached him to ask whether they could skip Phase I and go directly to Phase II. Seeing good arguments on both sides, he said that the issue should be examined.
- Set-aside percentage: While SBA would like to see the percentage of R&D funds reserved for SBIR go up, it would prefer to see the total federal R&D budget go up.
- Outreach to the underserved: This is a major issue in the SBIR program as it is in other programs run by SBA, which believes that parts of the small business community—especially women and minorities—have not traditionally been reached. The agency, Mr. Hill said, does "not buy into the argument that there are not enough women technology owners out there to fill this market void."
- Administrative costs: SBA would argue against allowing administrative costs to be part of SBIR, believing that the program's budget for awards is small enough without "taking more money from small business and giving it to bureaucrats to run the program." The question is a relevant one, however, and SBA is soliciting the opinions of concerned parties.

BALLISTIC MISSILE DEFENSE ORGANIZATION

Carl Nelson
Program Manager (ret.)

The Question of Subsidies

Dr. Nelson, for ten years head of the SBIR program at the Ballistic Missile Defense Organization (BMDO) and its predecessor, the Strategic Defense Initiative Organization, called into question at the outset the notion that the government can create wealth by granting subsidies. The standing controversy surrounding direct government subsidies is only heightened at a time when the private venture-capital sector is investing as heavily as it did in 1996 and 1997—when it provided \$10 billion and \$11 billion, respectively—and when angel investors are putting up many times that amount. At present, there is little prospect that these levels of investment will change. Annual return on venture capital has been at about 40 percent for the past four or five years; until the level of returns declines, venture money keeps getting recycled and will thus continue to pour in.

Although the market is concerned with return on investment and total return, most SBIR managers and overseers focus on scientific and technical merit. The mission agencies in particular, needing high technology, focus on performance rather than on market aspects or cost. The government does not live in a market world of instant information and instant response—nor would that be appropriate, Dr. Nelson said, reminding the audience that "one thing you want from government is stability." But, attributing "veto power" to the government's internal processes, he warned: "You can not get much done in a federal agency without a consensus of at least half a dozen people, any one of whom can say 'no' and then your proposition is effectively dead."

Improving SBIR: Incentives for Program Managers

Dr. Nelson put forward suggestions for two ways that SBIR might be improved under current legislation, which gives the participating agencies ample operating room to run versions of the program suited to their needs. Each agency should be allowed to either choose its own "empowered decision makers" for the program, or—if it is reluctant to move away from committees and place responsibility for decisions on individuals—adopt a formula, which might resemble the "ingenious" Fast Track pilot instituted by the DoD.

Basing a model of the first option on his own experience running the SBIR program at BMDO, Dr. Nelson likened the decision maker to an investment portfolio manager in contrast to a technical committee or branch chief, such as those at the Wright-Patterson or Naval Research laboratories. This decision maker would, within general guidelines provided by the agency head, have the authority to make awards and would, as a result, be held accountable by "some type of objective measure." Incentive would be created by providing a reward—perhaps a bonus of $10,000 each for the government's three best empowered SBIR managers in a given year.

To judge which managers are the best, some external objective standard would have to be established. Among the possibilities:

Return on investment. This is the typical standard in the venture-capital community, but some rethinking would be required to decide what the investment and return were in the context of a government agency. In addition, there is a problem of the time scale.

Public capital raised. Calling the amount of public capital raised, particularly through initial public offerings, "an excellent measure," Dr. Nelson stated as one of his principles: "If you are not heading toward getting public capital into a technology that's going to become a business [and thus] have real impact, . . . the scale of your thinking is probably too small."

Median size of awardee company. Although not necessarily a very good measure of results, this may—in light of the number of studies showing that the most innovation per employee comes from firms with under 20 employees—be an indicator of whether funding is going to the most innovative part of the small-business community. Using a control group of nonawardee firms, a technique pioneered by Dr. Lerner, would be integral to establishing this as a metric.

"Honor Roll" Selections. Another possible metric would consist in counting the number of firms in each manager's portfolio that appears on an honor roll of the type published by *Forbes*, *Inc.*, *BusinessWeek*, and Deloitte & Touche.

Improving SBIR: A Formula Approach

The second option, the formula approach, would put a ceiling on the percentage of an agency's annual SBIR budget that could go to the ensemble of firms that, on an individual basis, already held grants valued at a specified minimum amount; once that ceiling had been reached, these firms would have to compete against one another in a restricted pool. In addition, the ratio of matching funds to SBIR grant funds demanded of an award winner would escalate with the total amount of government money the firm had received under the program. Conversely, a minimum percentage of funding would be reserved for award to first-time applicants.

Limiting the availability of funding to established SBIR winners would correct the current tendency of committees and agency technology heads to award a disproportionate amount of their available money to firms whose technologies further their particular interests. This tendency appears to go hand in hand with the choice of narrow topics for SBIR competitions, which Dr. Nelson himself avoided in favor of broad topics. Because agencies may receive "a large number of proposals for a narrow technology, only one or two of which may be really competitive," he contended, a narrow focus is "almost arbitrary from the standpoint of commercialization efficiency." Similarly, in concentrating awards in few hands, program officials "are ignoring the power of the commercial market to develop a technology so that the agency can use it."

Requiring that financial contributions by awardee firms escalate with successive awards, as well as setting aside funding for first-time awardees, would be in line with Dr. Nelson's own past practice. At BMDO, however, he based his demand for a rising ratio of matching funds not on the total number of SBIR awards won by a firm but on the number of SBIR awards it had won in support of a particular technology. "I would say that if, by the time you got to the third award, you did not have a large ratio of matching share or money of some sort," he recalled, "there was something wrong with your underlying assumptions about the marketability of your technology." The government's selection process was helping foster what he charged to be a "large measure of self-deception going on" within the SBIR program. Granting Phase I money to first-time applicants—which he did in the case of almost any new bidder with an "innovative-sounding idea that at least the technical experts did not say violated the Second Law of Thermodynamics"—and instituting schemes such as Fast Track that are designed to demonstrate private-sector interest in a technology are essential, in his opinion, "if you are going to get a wheel invented—not reinvented."

PARALLAX RESEARCH, INC.

David O'Hara
President

Mr. O'Hara began by emphasizing the record of nongovernment sales established by his company, a maker of x-ray optics included in x-ray analytical instruments, primarily for the semiconductor industry. Parallax got its start in 1994 with the help of a SBIR award from BMDO. Sales to nongovernment customers accounted for only a few percent of gross revenues in 1995, but that percentage climbed to about 10 percent in 1996 and about 25 percent in 1997. It is projected to reach somewhere on the order of 60 percent for 1998.

Personal Perspectives on SBIR

Although this record contrasts sharply with that of the company that employed Mr. O'Hara before he founded Parallax Research, it was shaped in large part by the perspective he gained working there. His former employer, also a small firm, had repeated success in obtaining SBIR awards—the proposals for many of which were written by Mr. O'Hara himself—but was never able to commercialize. This company eventually received a letter from Dr. Nelson accompanying a Phase II award stating that before receiving a $100,000 award, the company would be required to present significant evidence of commercialization in the form either of an agreement from a customer to buy the product or of matching funds.

Because the firm's president was "horrified" by this stipulation, a meeting was arranged with Dr. Nelson who explained that one of the better ways of evaluating the commercial prospects for an SBIR award was gauging the project's ability to attract "other people's money." Investment by the awardee firm itself was less significant than that by an outside company, because a third party is unlikely to invest in something that it thinks will be a loser. "My company president had worked all his life as a government contractor, and this did not make any sense to him," Mr. O'Hara recalled. "Looking at him and seeing that he just did not get it, I realized that I was in the wrong corporate culture." After observing at a trade show several months later that his boss was very interested in sessions at which scientific papers were presented "but showed no interest whatsoever in walking among the vendors and getting his hands dirty," he decided to leave the firm.

Within six weeks of forming Parallax to commercialize some of the technology that had been developed by his previous employer, Mr. O'Hara obtained a large purchase order for a product. He was not sure, however, whether the product could actually be produced; in fact, he had been told by staff at NASA and at the Lawrence Berkeley National Laboratory that production probably was not feasible. Using liquidated personal assets to operate Parallax, he achieved technical success in the ensuing months but was beginning to run out of money. At that point, his firm was saved by a Phase I SBIR award from BMDO, at the end of which it was able to build a prototype and to demonstrate that the technology "really worked."

Commercialization and Phase II

When the firm applied for a Phase II award, Dr. Nelson asked Parallax, as he had Mr. O'Hara's previous employer, to show significant evidence of its technology's commercial potential. A first indication was the purchase by a larger company of the firm with which Parallax had been collaborating, corroborated by a statement from the new owner that it had made the purchase solely because of its enthusiasm for the instrument that was being developed on the basis of Parallax's technology. Then when this company put up matching funds, Parallax was able to get the Phase II award. Parallax has developed three products based on its technology, the first of which was introduced in the summer of 1997 and the second in September 1997; a third is scheduled for introduction in the spring of 1998. Because the firm has not come to the end of its SBIR funding, it is planning the introduction of a fourth product as well.

Mr. O'Hara stated he agreed with the premise that firms will not have much difficulty obtaining matching funds when they have good ideas that look like they are headed for commercial success; "I've given it one shot," he said, "and it's worked." He has received offers to buy not only the product that Parallax is on the verge of introducing but the firm itself. Still, he has reservations about the fact that the period that follows Phase II but precedes commercialization may turn out to be a long one. When it comes to the schedule for commercialization, larger organizations typically have a much longer time scale than a small, start-up company like Parallax, something Mr. O'Hara believes the funding agencies should take into account.

ADVANCED TECHNOLOGY MATERIALS, INC.

Gene Banucci
President

While remarking that Advanced Technology Materials, Inc. (ATMI), is a few years further down the road than Parallax, Dr. Banucci said that the story of his firm's success under SBIR compares with that of Mr. O'Hara's. After going into business in late 1986, ATMI won a SBIR grant from the Department of Energy (DoE); the product it yielded was promptly commercialized and, having booked almost $60 million in revenues to date, went on to become the largest single success in the history of DoE's SBIR program. ATMI's next SBIR award, from the Environmental Protection Agency, also resulted in prompt commercialization; total sales of the product developed from that award are approaching $100 million, while the product line based on it had brought in $30 million in revenues in the first six weeks of 1998. Dr. Banucci expects what he believes will be the program's first billion-dollar product to result from ATMI's subsequent SBIRs: Separate awards from BMDO and NSF, both of which spanned Phase I and Phase II, helped with different aspects of the research.

ATMI's Business

ATMI went public in 1993; its market capitalization, initially at $35 million, now exceeds $600 million. What he called "a SBIR partnering strategy," which includes working with major corporations and integrating them into ATMI's business plan, has been the key to the firm's growth: "We took the intense pressure that Carl [Nelson] applied with regard to partnering and getting matching funds to heed because we desperately needed that money to grow the business." Because the firm is virtually certain to add 100 employees to its current work force of 400 by the end of 1998, this year is likely to be the last in which it will be eligible for the SBIR program. "By any measure," he said, "ATMI has been a SBIR success story, and we are very proud of saying that this program has been the catalyst for making ATMI a success in the public market."

The company, which had revenues of around $102 million in 1997 and has facilities around the United States and in other countries as well, makes materials that go into a reactor to coat the silicon wafers from which semiconductors are made; it is these coatings that give each device its unique electrical and physical properties. ATMI has implemented a three-phase growth strategy, entering three niches: basic materials used in making semiconductors; environmental equipment used in managing those materials; and proprietary thinfilms, which it produces as a service business.

ATMI's Technology

The product developed from the research aided by the BMDO and NSF awards is designed to provide a safer and more environmentally benign alternative for storing and transporting toxic, corrosive, or otherwise hazardous gases used in the semiconductor industry, which are generally packaged in pressurized cylinders. The new technology involves placing a solid absorbent in a standard gas cylinder, into which gas is then introduced; in the sealed cylinder, the gas is converted to a solid by its absorption on the solid substance, and it thus remains there without having to be kept under pressure. The absorbent is designed to release the gas under vacuum, which fits with the manufacturing environment of the semiconductor industry where most processes take place at reduced pressure. Thus, a vacuum pump draws the gas off the absorbent and out of the cylinder into the reactor, rather than pressure inside the cylinder expelling the gas.

Not only are gases that are stored using ATMI's technology safer to transport and handle, but the technology offers an improvement in productivity as well. Although the amount of gas that can be put into a conventional cylinder is dictated by the degree of pressure the cylinder walls can take, converting the gas to a solid form removes that limitation; and, in any case, a given amount of a substance takes up far less space as a solid than as a gas. ATMI has been able to put between five and ten times as much gas in its cylinder as can be stored under high pressure in a cylinder of the same size, a result that promises to allow gas supply cylinders in semiconductor fabrication plants to be changed that many times less frequently. The potential for reducing downtime and lowering cost "means we can charge anything we want for that product," Dr. Banucci said. He expects earnings of the kind that made it possible for ATMI to return $6 million to the government last year in the form of taxes.

The Role of SBIR

He called SBIR "an absolutely terrific program" and urged the audience not to lose sight of how well it is working in general, even if some fine tuning needs to take place, as it must in "any program that is going to be at peak form." Endorsing Mr. O'Hara's view of the ease with which SBIR winners that have a unique technology can partner, he displayed a list of 13 major companies that have invested at least $1 million in a partnership with ATMI and stated that 85 percent of the money that has been put into the company has been related in some way to an SBIR award. To push SBIR recipients toward commercialization, the program could increase the size of awards to firms that have lined up partners willing to share costs. On the other hand, a mechanism should be sought that would put the "SBIR mills" out of business, and award winners must be held accountable for commercializing their ideas if the program is to achieve the goals advocated for it by Rep. Davis and Dr. Nelson.

AMERICAN XTAL TECHNOLOGY

Gary Young
Vice President

Placing his company between those of Mr. O'Hara and Dr. Banucci on the development continuum, Mr. Young said its business, making gallium arsenide substrates, is too small to interest either large companies such as AT&T—which abandoned the technology American Xtal Technology (AXT) has developed—or venture capitalists in the United States. In fact, the United States imports between 70 and 80 percent of its needs from Japan, which dominates the gallium arsenide substrate industry worldwide; because of the product's importance to electronics sectors downstream, large Japanese companies are willing to lose money year after year manufacturing gallium arsenide substrates. But while Sumitomo and Hitachi run what Mr. Young described as the equivalent of a "small DoD" to keep Japan's supply base intact, the Pentagon itself is obligated under Title III to ensure the survival of a domestic supply base for gallium arsenide substrates as a strategically important material.

AXT's Business Strategy

Mr. Young, who like Mr. O'Hara, has liquidated personal assets to finance his company, sees product commercialization as the only path to a financial return. AXT's business strategy stresses

- high-quality output;
- providing the customer both with security of supply and the best bottom-line cost;
- diversifying the product mix;

- selling both in the U.S. domestic market, which accounts for 60 percent of its revenues, and in markets abroad; and
- supporting R&D.

The Role of SBIR

As a result of adhering to this strategy, he claimed, AXT has had a "very smooth ride" and enjoys a favorable reputation in its industry. And, with the help of SBIR, the company—initially too small to qualify under Title III, which targets manufacturers already established in their industries—has proved itself to the government by delivering on its promises and has grown large enough to be accepted under Title III.

The capital demands resulting from AXT's growth rate, which has averaged about 60 percent per year, have added to the pressure to commercialize. The firm plans to commercialize products developed through the two SBIR awards it has received: One was for gallium arsenide technology, the second was for indium phosphide technology, the latter described by Mr. Young as a little more expensive and, as a material, a little faster than gallium arsenide technology but otherwise similar. Alluding to comments made earlier that if SBIR supports only winners, it is being too conservative in its awards, he added that AXT's 100 percent success rate for commercialization makes it a "bad example."

The market for gallium arsenide substrates, used in such telecommunications products as cellular phones and fiber-optic laser diodes, took off in 1995 thanks to an explosion in the popularity of the former product. At present, AXT has 230 employees and is the leading supplier of gallium arsenide substrates in the United States; its goal is to become world leader in 1998 or 1999. In the third quarter of 1996, it built a 50,000-square-foot production facility on 4.6 acres of land it had purchased the previous year. A 30,000-square-foot expansion is to be completed in either February or March 1998, and the company is looking to open another site, in the range of 50,000–100,000 square feet, in 1999. Profitable for 28 consecutive quarters, AXT has paid more than $8 million in taxes, showing its proprietors to be either "good citizens or bad tax planners." The company recently completed filings with the Securities and Exchange Commission.

DISCUSSION

The Need for Cooperation

Terry Bibbens of the SBA began by praising the National Research Council for convening members of the academic, small business, and SBIR communities and by urging the various communities to work together: "When the Army went up and battled the Navy for appropriations and vice versa," he recalled, "neither one of them got funding." Then, explaining that it is the job of SBA's Office of

Advocacy, where he serves as Entrepreneur-in-Residence, to ensure that federal rules and regulations have no unfair negative impact on small business, he argued that obliging SBIR award recipients to put up matching funds would in fact be unfair.

Matching requirements should be attached to SBIR awards only if identical requirements are placed on the remaining 97.5 percent of federal R&D funds, which generally go to larger companies and universities. "Any singling out of small business to match funds for the research agenda of this nation is unfair and unreasonable and penalizes small business," he stated. Pointing out that during the current economic recovery, small business has created 12 million new jobs while 2 million jobs have been lost by big business, he warned policymakers not to "put hurdles in front of the small business community."

Furthermore, imposing matching requirements on SBIR awards would be unfair in light of the venture-capital industry's aversion to small deals and of the absence of a model whereby small firms can line up matching funds from other sources. Fewer than 3,000 companies annually receive investment money from the U.S. venture community, including from Small Business Investment Corporation, and the average size of those investments is now approaching $13 million. "The massive amount of funds that come into the venture community does not even apply to the SBIR community," stated Mr. Bibbens, adding that "SBIR companies are not looking for $13 million, they are looking for a million, a half-million, maybe two million [dollars]." Applauding the DoD's Jon Baron for the Fast Track pilot program, he expressed the need for a vehicle that would draw investors to companies of the size that apply for SBIR awards.

Focus on Commercialization

Responding to Mr. Bibbens's first point, Dr. Nelson placed SBIR moneys in a different category from the other 97.5 percent of the federal R&D budget on the grounds that the "objective of commercialization" stated in the law creating SBIR is "unique for R&D in the government." Because the amount of money going to the small business community is set under the law, small firms are not competing with large institutions in any way. Moreover, he pointed out, he had never required matching funds; rather, he had told proposers that he would make the existence of matching funds "a live, competitive factor" in selecting Phase II awards. But although never a requirement, matching funds, often on a dollar-for-dollar basis, started showing up in Phase II proposals beginning around 1995 because the proposing companies, realizing that commercialization had been made a criterion, "could not wait to go out and prove their commercialization merit by bringing in commercial partners."

Only about 20 percent of this matching fund came as money from venture-capital firms or angel investors; the remaining 80 percent was from third-party firms, and a large part of that was in-kind aid provided within their own companies in some way that would advance the technology's position in the marketplace. There was no contractual requirement regarding what had to be done by the SBIR winner's partner company; marketing, testing, and R&D were all acceptable. "My only requirement," said Dr. Nelson, "was that the contracting officer saw that the money was being spent in furtherance of the technology."

Panel V

Improving Assessment and Selection

INTRODUCTION

Dan Roos
Massachusetts Institute of Technology

Dr. Roos opened the session with the announcement that Herbert Kreitman would substitute for Wendy Baldwin as the representative of the National Institutes of Health (NIH) on the panel. He then introduced Maryellen Kelley, serving as a senior economist for the Advanced Technology Program (ATP) at the National Institute of Standards and Technology while on leave from her post as associate professor at the Heinz School of Public Policy and Management at Carnegie Mellon University. Dr. Kelley discussed the lessons emerging from ATP from two perspectives: (1) to what extent small businesses are involved in the program, and (2) how ATP compares with the SBIR program from the viewpoint of assessment, selection, and objectives.

LESSONS FROM THE ADVANCED TECHNOLOGY PROGRAM

Maryellen R. Kelley
NIST and Carnegie Mellon University

Differences between ATP and SBIR

Beginning with a description of ATP's operation, Dr. Kelley outlined features of the program that make it unique in the federal government's technology policy portfolio. ATP is aimed at directly supporting R&D activities that have a commercial orientation from their inception, unlike SBIR which pursues the goal of commercialization through R&D that supports the mission of the SBIR-member agency making the award. Like SBIR, as well as most government agencies that are engaged in R&D, ATP funds activities that are undertaken by private companies; in ATP's case, a cost-sharing requirement is designed to leverage private R&D investment with the goal of achieving public benefits.

ATP is intended to support "enabling" technologies, which are characterized by both high technical risk and the potential for substantial, widespread commercial application. Technologies considered enabling can be:

- **path-breaking**—those that promise to induce radical innovations;
- **infrastructural**—those that support R&D, production, or businesses in entire industries; or
- **multi-use**—those that either become platforms for multiple paths to further technology development or are "generic" in that they have many applications.

Rather than advancing a specific agency mission, the program's endeavors are "industry driven." An annual general competition is open to proposals from all fields of technology; in addition, there are numerous competitions each year devoted to preselected areas of technology focus that have been identified by ATP program managers in cooperation with industry.

Evaluating ATP

Study of the impact of the program, which was inaugurated in 1990, has been limited to date because of the long-term nature of the projects it funds. In the early stages, benefits that become apparent are largely in the generation of knowledge—knowledge that may, to some extent, also be of use to non-ATP firms and for technologies other than those directly targeted by an award. In the long term, knowledge spillovers are expected, as are other types of benefits that do not accrue exclusively to those involved directly in the innovation process. In the case of infrastructure-enabling technologies, other firms may learn about the technologies developed within ATP and build upon them. Also envisioned are market spillovers: Knowledge that reaches beyond the ATP-awardee firms may allow other firms to build cheaper and better products, eventually benefiting the consumer.

"These are the ways in which the larger benefits of the program are realized in the economy," explained Dr. Kelley. "That, in some sense, is the major justification for this type of effort: that it will have benefits beyond simply the private returns to the individual firms."

Implications of ATP Evaluation for SBIR

ATP sees monitoring and evaluation as integral to an ongoing process of refining the program so that spillover benefits can be captured more effectively. Tracking projects as they progress, ATP's management is able to consider implications for the program's strategy while at the same time building a database for use in its long-term evaluation. Proposing that ATP's experience might be of benefit to SBIR, Dr. Kelley provided a description of its evaluation methods:

- Keeping a database on projects and applicants "from the very beginning" allows ATP to track the characteristics of the applicants, projects, and types of technologies it funds.
- Progress on individual awards is charted in two ways:
  - — Participants are required to fill out documents for ATP's internal use; agreeing to provide information and to be studied is, she said, "in some sense the quid pro quo for being an awardee of the ATP." These documents help ATP staff gain knowledge about both the effectiveness of the program and the issues that arise in conducting it that may contribute to the quality of program management.
  - — R&D activity is monitored by ATP project managers with the help of staff visits to participating companies and technical reports that awardees are required to provide on a quarterly and annual basis.
- Studies are commissioned by ATP that may focus on a particular technology area, a particular industry, or even particular firms, and that may involve projections of the long-term benefits to be expected from certain types of awards.
- Methodological research questions are explored in the field of metrics, addressing the question: "What do we need to know to be able to determine whether or not a project is successful?"
- Projects are assessed for their ability to meet the goal of providing broad returns to the economy at large. Social benefits are not studied to the exclusion of private returns, however, because the latter "are, after all, part of the social return."

Because ATP's funding did not reach its present order of magnitude until 1994, and because most of the awards have been to multiyear projects, Dr. Kelley noted, very few projects have been completed. The program must be considered as still at a very early point in the process.

Turning to data analyzed so far, Dr. Kelley said that 3,000 proposals from a wide area of technological activities have been submitted over the life of the program. These proposals have requested, in total, far more money than has been available in ATP's budget, and therefore only a small percentage has been awarded. Whether the proposals selected have been efforts undertaken by individual companies or collaborative joint ventures involving groups of companies, all have involved some degree of cost-sharing on the part of recipients. And although ATP, far from targeting small firms, has large and small firms competing for awards, more than half of the projects funded under the program to date have been to small firms or to consortia that had small firms as their leaders.

THE SBIR EXPERIENCE AT NIH

Herbert Kreitman
National Institutes of Health

Dr. Kreitman, the coordinator of the SBIR and the Small Business Technology Transfer (STTR) programs for the U.S. Public Health Service and NIH, opened by noting that NIH's budget has grown significantly over the past few years to $13.6 billion in 1998. The Clinton administration has proposed that it surpass $20 billion in 2003. In the current year, NIH SBIR and STTR set-asides total $281 million, keeping the agency in second place among all SBIR agencies behind the DoD.

Whether defined as sales within the federal government or to the private sector, commercialization receives less attention from NIH than from most other SBIR-participant agencies. Although commercialization is among NIH's review criteria, more than 95 percent of the agency's SBIR awards are made as grants, for which review criteria are weighted differently than when awarding contracts. Furthermore, Mr. Kreitman suggested that the metric for commercialization be expanded to include published papers describing the results of SBIR projects and the use of SBIR technology in other research projects. Despite all this, he noted that in the most recent U.S. Government Accounting Office (GAO) report on SBIR commercialization, which dates to 1992, NIH was "held as the standard for commercialization."

Describing other characteristics of the NIH SBIR program, he said that there is no topic number on the face page of the agency's application form. It is the investigator-initiated research project that NIH is most interested in, and any project that fits its mission of biological and behavioral research is welcome. The agency has a Fast Track program that is probably somewhat more flexible than the DoD pilot. NIH's Fast Track, while it encourages matching in the amount of a Phase II grant, does not require matching at any specified level and accepts it in the form of resources as well as cash. NIH's Fast Track program also requires a product development plan, which is not a requirement of its standard Phase I or Phase II applications.

Mr. Kreitman concluded by noting that, although Congress has treated NIH generously overall, it has placed restrictions on the agency's research and management support budget line. NIH believes it could do a much better job of nurturing the small business community if it had additional resources for outreach. Even though the small business community must be represented on the agency's SBIR and STTR peer-review panels, the majority of reviewers come from academia. "So, yes, there is grantsmanship. You have to, as best you can, get into the heads of the reviewers," he said, adding: "We are perfectly willing to provide such nurturing, but we need some resources to do that."

THE FAST TRACK PILOT

Jon Baron
U.S. Department of Defense

Mr. Baron specified at the outset that the central thrust of reforms made in the DoD's SBIR program over the past two years has been to increase the program's success in converting the research it funds into viable products that would make a major difference to U.S. military capabilities. The reforms, of which Fast Track is one, grew out of an assessment done in 1995 by Dr. Davis's office within the Office of the Director of Defense Research and Evaluation (DDR&E). Dr. Flamm, who headed DoD's dual-use technology office at the time, and Dr. Robyn of the White House National Economic Council were involved as well.

A process action team comprising representatives of various DoD agencies—including the program managers for SBIR from the Army, Navy, Air Force, and defense agencies—developed the actual reforms. Although sales of products from SBIR grantees to customers outside the department economy were also goals of the reforms, Mr. Baron emphasized that the reforms' major objective was to benefit the DoD itself in a significant way.

Overview of DoD's SBIR Program

The funding level for SBIR in DoD during FY 1997 was $547 million. This represents not only half of the government's entire SBIR program, but also one-fourth of all federal R&D funds that go to small business. The Pentagon's SBIR program is administered by eight component programs within overall policy guidance from the Office of the Secretary of Defense. A 1997 GAO report focusing specifically on DoD's SBIR program was "quite positive," Mr. Baron said, in line with the consistently favorable reviews accorded the SBIR program as a whole in a number of independent studies.

In addition, a study funded by Dr. Davis's office at DDR&E found that Phase II awards under DoD's program between 1984 and 1992 yielded average product sales in Phase III of $760,000. That study virtually replicated a 1992 GAO study of commercialization, which had placed average product sales in Phase III at $285,000; the substantial increase presumably stems from the fact that the technologies considered have had more time to mature into actual products.

As an example of the program's success, Mr. Baron named Viasat, an SBIR start-up company funded by both the Air Force and the Navy that developed a technology used in military communications satellites. Before the implementation of the technology, each user of a military communications satellite had to set up a dedicated channel. This would tie up the satellite for a certain period of time during which no others could use it. By obviating the necessity of a user's tying up a satellite other than when it is actually communicating, Viasat's technology has effectively enabled military communications satellites to handle up to ten times as many users in the course of a day. Sales to DoD now stand at about $90 million, and the technology has begun generating significant sales to the private sector for commercial communications satellites. The commercial market promises to exceed $100 million; orders from AT&T and others have already surpassed $7 million.

DoD's Fast Track Pilot

In 1995 the DoD process action team made a careful review of the department's SBIR program, looked at previous studies by GAO and others, and spoke with a number of companies that had participated in the program—in particular, but not exclusively, companies like Dr. Banucci's that had become "huge" successes. Based on that review, the team came to a conclusion very similar to that of both the DDR&E and GAO the studies: Although the program was very successful, the success was concentrated, with the top 1.5 percent of Phase II award winners accounting for over 50 percent of that success as measured in Phase III sales.

In particular, the team concluded that many of the companies taking part in the SBIR program are very strong in R&D capability. However, they are weak in the capabilities that would translate their R&D into real products that could be sold to DoD. Companies that had won ten or more SBIR awards in the period 1984–1992 had on average about 40 percent of the sales resulting to both DoD and the private sector. "These are companies that write good proposals [and] do good research and development," Mr. Baron observed. "Their research and development has had some benefit to the department, but not the same significant benefit as if [it had been] converted into actual products that were sold to DoD in significant numbers and incorporated into defense systems."

Among the reforms put forward by the team to address this problem was the SBIR Fast Track pilot project. Under Fast Track, SBIR projects that attract "a little bit" of matching cash from outside investors toward the end of Phase I gain a significantly higher chance of obtaining a Phase II award; they also receive expedited processing and interim funding that, together, amount to continuous funding between Phase I and Phase II. As explained by Mr. Baron, the rationale for this policy is that if a company attracts matching money—whether from a venture-capital firm, defense contractor, or other company, or by meshing with an acquisition program within DoD—this sends an important signal. "This says more about the size of the market for its technology and about the company's ability to bring that technology to market," said Mr. Baron, "than anything the company could write down on a piece of paper in a proposal."

Although it is still in the pilot stage, Mr. Baron said that results so far show that, in general, Fast Track has been meeting its intended goals. Last year, with the help of DDR&E and the managers of the component SBIR programs at DoD, the pilot was streamlined and some administrative problems were corrected, whereupon it was extended for two years. Scheduled for next year is an independent, systematic evaluation of Fast Track that is designed to determine whether the favorable early results have been sustained and whether the improvements have accomplished their objectives. Anecdotal evidence so far indicates that Fast Track has provided an effective means for companies to leverage their SBIR awards to obtain outside funding. This is because, as Mr. Baron explained, Fast Track has allowed them to entice potential partners with the prospect of their investments' being matched at ratios of between one-to-one and four-to-one by the Defense Department.

Returning to the DDR&E study, Mr. Baron noted that private-sector investment in SBIR projects seems to correlate strongly with whether the awardee will have success in generating sales of new products. According to the study, SBIR awardees that had failed to attract any private-sector investment had about a 2 percent chance of achieving sales a few years down the line. For those that had attracted more than $1 million in outside investment, that chance rose to about 40 percent. This holds for a subset of SBIR awardees in which sales are defined solely as sales to the DoD. If an SBIR project had no private-sector investment, it had about a 3 percent chance of achieving sales to DoD or to DoD prime contractors in Phase III, a figure that went as high as 30 percent for awardees attracting outside capital. This evidence, he said, very strongly suggests that, contrary to claims that conflict exists between agency missions and the goal of private-sector commercialization, the two objectives are in fact complementary. The central challenge is getting the R&D converted into saleable products, regardless of who the customer is.

Other DoD SBIR Reforms

Mr. Baron concluded what he characterized as a "very preliminary" report by enumerating other changes enacted as a result of proposals by DoD's SBIR process action team:

Phase II declaration: A company submitting an SBIR proposal must now list all Phase II awards it has won, as well as declare the sales revenues stemming from those prior Phase II awards and the amount of non-SBIR money it has raised to match them. The Ballistic Missile Defense Organization and some other organizations use this practice in their evaluation processes; at this point, it is unclear what effect this has had.

Streamlining: DoD has made an effort to streamline the processing of SBIRs across the department, and thanks to the work of the component programs there has been a 30 percent reduction in funding delays. Because time to market is critical in most high-technology industries, the year's delay between Phase I and Phase II that the department experienced in the past had biased the program against success in the private sector.

Assessment: Prospective SBIR proposers can now talk with the DoD scientists and engineers who author the program topics before submission. This innovation, which originated in the Air Force and was also recommended by DDR&E, allows companies that have never done business with the department to get a much better sense of its needs before they submit a proposal.

DISCUSSANT

Charles Rowe
House Committee on Small Business

Claiming he lacks the qualifications to comment on "a selection and peer-review process that's placed mostly in the hands of people with Ph.D.s," Mr. Rowe indicated that his remarks would focus on the results of the SBIR program. These results, as appraised by the National Academy of Sciences, GAO, or the Small Business Administration, have shown the program to have been effective over the years. "Although the vast majority of federal R&D funding goes to universities, large businesses, and federal laboratories, we see that the vast majority of patent applications are coming out of the small business end of the pipeline," he stated. "That is only to be expected from a program that is fundamentally based on competition, merit, and commercialization. There is not much reason to patent something that you are not planning to sell."

Distribution of Awards

What has been termed a "lopsided distribution of awards" is the largest single problem with the selection process that has been brought to the committee's attention over the years. Calling this phenomenon "both a boon and a curse to the program," Mr. Rowe signaled his approval of the fact that awards have been concentrated in certain geographical areas. Pointing out that, according to the study presented by Mr. Tibbetts, award winners are usually located in "places where the money is," he argued that the presence of money aids Phase III commercialization. Calling the SBIR program competitive and the proposals it attracts excellent, he said that the committee has yet to see a single concrete example of an award that was unworthy of funding, while a desire to find money to fund more of the proposals has been evident.

The Phase I and Phase II Funding Gap

A second and far more concrete complaint coming out of the selection process involves the gap between Phase I and Phase II funding that DoD's Fast Track program is designed to remove. Mr. Rowe expressed his pleasure at noting that NIH has put a version of Fast Track in place as well.

Although there may be problems with some versions of the selection process at the 11 agencies that participate in SBIR, Congress in creating the program had very deliberately kept its fingers out of the management of the selection process. Leaving it to the agencies themselves creates a spirit of competition among them that will be far more productive in the long run than micromanagement from Capitol Hill.

DISCUSSION

Ann Eskesen of the Innovative Development Institute stressed the importance of Dr. Nelson's observation at a previous session that the SBIR program's framework allowed him the latitude to design his own program. This allowed him to get to know the companies participating in it extremely well. Similarly, she underscored Mr. Kreitman's point that an agency running a program on the scale of that at NIH needs the resources to stay in contact with the companies involved. "There very clearly is a correlation between the effectiveness of participation in the SBIR program and the amount of interaction that occurs," she stated, "whether it occurs within the agencies, to the companies, or within those companies and the state support organizations."

Looking forward to SBIR's reauthorization, Ms. Eskesen advocated moving beyond the outreach effort that she views as having been the program's emphasis since its creation in 1982. Focus on getting the small firms to participate in the program has, to date, overshadowed concern for Phase III, during which it was simply "assumed that something was going to happen." Although "for the really good companies it did happen," the lesson of the past few years is that some companies need support and access to resources rather than just the incentive to be involved. Because SBIR is now a mature program, and because the environment in which it is functioning is fundamentally different from that in which it was created, the needs of awardee companies merit consideration as the reauthorization approaches. Referring to the difficulty many companies have in bridging the "funding gap" between R&D and demonstration, she said that "setting standards and just pushing the bar up higher and higher works for some companies but is not working for all."

Ms. Eskesen then turned to the general question of how the integrity of the SBIR program is to be maintained in light of its sheer diversity of involvement. Declaring that one size does not fit all in the case either of the agencies or of the companies involved, she urged an examination of how micro-management is to be avoided and individual program managers left with the freedom to experiment. As the program has become institutionalized with age, and as its rising funding level has brought increased visibility, this freedom has diminished. "With all due respect to the present set of program managers," she stated, "a lot of your predecessors made career-threatening decisions. They got away with it because the SBIR program was small and somewhat in the corner, and most people did not even know it was there."

Stressing the value of interaction, Ms. Eskesen observed that managers in some agencies have little or no contact with the companies in their awards portfolios. Technical officers she observed at NASA flight centers, for example, "would not have known their companies if they walked through the door." They simply had no resources that would have allowed them to get to know the firms. Although the SBIR program works in its present form, improvements are necessary if it is to fulfill its potential, Ms. Eskesen stated. She invited the panelists to name one or two changes that might make a difference to SBIR personnel who interact with companies and make the operational decisions.

Panel VI

Observations and Policy Issues: Agency Perspectives

INTRODUCTION

Charles W. Wessner
National Research Council

The purpose of the day's concluding panel, Dr. Wessner said, was to provide SBIR program managers an opportunity to comment on the proceedings. He put forward several questions:

- What have the managers found most helpful?
- What have they disagreed with most strongly?
- Are there features of their own programs they regard as key?
- Have they gained any new insights?

Arlene de Blanc
U.S. Department of Energy

Tensions with the SBIR Program

Ms. de Blanc introduced herself as one of three people who manage the SBIR program at the Department of Energy (DoE)—which at about $75 million annually is the fourth largest in the government. Ms. de Blanc said that the most difficult problem she has seen in her 18 months with DoE's program is the tension between the goals of the agency program offices and the academic R&D

community on the one hand and those of the SBIR program on the other. Invoking a scene from Lewis Carroll's *Through the Looking Glass*, she recalled that when Alice, fresh from stepping through the mirror, reacts to the White Queen's disarray by tidying her up, the monarch is so impressed that she offers the young girl a job. When the White Queen attempts to persuade the reluctant Alice to accept by offering to pay her wages in jam, Alice replies: "I do not care for any jam today, thank you." At this point, the White Queen becomes irate, telling Alice: "You could not have it today anyway, because we have jam every other day. We had jam yesterday and we have jam tomorrow, but never jam today."

This, declared Ms. de Blanc, parallels what she saw in the DoE research community. The researchers were willing to recognize SBIR's past history of substantial and very positive economic and social results—"that's our 'jam yesterday.'" They were also willing to recognize that there would be future fruits of commercialization in the form of spinoffs—"jam tomorrow." But, they asked, "where is the contribution to our research program today—where's our 'jam today'?"

Over more than a decade, Ms. de Blanc observed, DoE was not particularly good at providing "jam today," having been focused on the other two days. As evidence, she pointed to the recommendation of a process improvement team, formed by the Department in response to discontent over the issue, that program research offices be given the ability to select their own projects. Reading the team's report, she had been "dumbfounded" to learn that the department had for years collected the SBIR "tax" from the research programs without permitting them to select projects. In the past year and a half, however, program managers have been able to select their own projects, with the result that the projects have been much more relevant to their programs than had someone else selected them. In addition, the program offices are virtually guaranteed to get back very close to the total dollar amount that they put in.

"All of a sudden, we see less tension between the SBIR program and the R&D community in DoE," she remarked. "All of a sudden, we see cooperation and joint efforts between SBIR projects and other non-SBIR projects as programs realize that they can further leverage their funds through partnerships with other organizations that are similarly oriented." She then pointed to a first for the department: a new partnership in carbon management under which three DoE research offices—fossil energy, biological research systems, and materials—are collaborating on one SBIR topic. "In order to get the 'jam today,'" she said, "we have to make sure that these constituent communities have a real stake" in the SBIR program.

Commercialization

Ms. de Blanc recalled Dr. Morgenthaler's suggestion that Phase II applications be evaluated for commercial potential. She noted that DoE has made such evaluations for 15 years, with sales, cost-sharing, and follow-on funding commitments making up a good percentage of a proposal's score. But that also has been a source of tension in the research community, which looks at research rather than commercial potential. The department has therefore created a scoring structure under which commercial potential, although important, is not controlling. "A research project that has no commercial potential can still make it through," she stated, "but it had better be very good."

Ms. de Blanc emphasized what she sees as the importance to SBIR of "matchmakers," without whom the program will be unable to "get all of [its] little mom and pop groups to marry off their progeny." DoE, realizing that one size does not fit all, this year has provided two very different matchmakers to address the needs of projects that have gone through Phase I and Phase II and are now looking into Phase III. She speculated that results of these efforts would be available in two years' time. Finally, she promised that if Congress will define what SBIR is supposed to be—"whether it is high risk, low risk, high R&D risk, low R&D risk, grant, contract, domestic, foreign, or all of these things together"— then the program managers will find a way to deliver "jam today."

Kenneth Gabriel
U.S. Army

The Opportunity to Expand SBIR's Impact

Pointing to SBIR's overall budget of $1.1 billion and to the fact that hundreds of smaller organizations are represented by the agencies that participate in it directly, Dr. Gabriel observed that the more latitude allowed to constituent organizations, the more difficult it becomes to administer a program that is explicitly focused on a given objective. In light of the divisions that separate government, industry, and academia from one another, he suggested that the most powerful benefit that the SBIR and Small Business Technology Transfer (STTR) programs could deliver might be in bringing the academic world closer to industry, both large and small. Large businesses can fulfill the infrastructural requirements, and academic institutions offer knowledge. "I believe that the biggest opportunity we have," he said, "is to make SBIR and STTR available not only to small business but to the entire ensemble of the marketplace."

The Role of Entrepreneurs

The entrepreneurial engine being provided by small firms has the potential of introducing new ideas into large corporations, while many growth ideas are being spawned at research institutions like MIT and in the complex that surrounds it. The efforts of the Army, Dr. Gabriel said, are in the direction of enhancing the participation of large businesses and academic institutions in developing technologies and in getting them successfully to the marketplace, so that small business does not carry the burden of doing the entire work of innovation alone. It should bè possible to make the SBIR program much more suited to these purposes than it is now; intellectual property rights are of concern if small businesses are to interact with large businesses with the assurance that such legal issues as transfer are covered.

Concluding, Dr. Gabriel named two crucial aspects of the Army's efforts:

- **Technical merit.** "We will fall on our sword on the technical merit" of an award, he stated.
- **Decentralization.** To ensure the broadest possible participation in the program, the Army allows final decisions on proposals to be made at its local facilities, "at arm's length from the collusion and 'neighborhood culture' that grows among the small businesses."

Robert Norwood
National Aeronautics and Space Administration

Dr. Norwood, who is responsible for SBIR at NASA as well as for the agency's commercialization programs overall, agreed with Ms. Eskesen's comments about the desirability of program officers' interacting with awardees. He said, however, that the decline in levels of both personnel and funding in the federal government makes it hard to see how SBIR would be able to support activities comparable to those that venture capitalists or investment bankers take in the process of due diligence to increase the chance of successful commercialization.

Metrics

Dr. Norwood expressed doubt that any one measure of success could prove useful for the SBIR programs of all 11 participating agencies, because their approaches vary so widely. He cautioned against relying on "success stories" of individual SBIR award recipients; doing so, he pointed out, assumes the validity of a cause-and-effect link between program support and business success that has not been well enough established to support an overall metric.

Increasing the Flow of Venture Capital

A potential route to improving the technology coming out of the program, Dr. Norwood suggested, might lie in determining whether private resources such as venture-capital funds can be brought in at an early stage of an SBIR project.

This must be done in such a way that there is no intrusion on the awarding agency's mission. He asked to hear from anyone in the audience who knew how to accomplish this within the terms of the law.

To conclude, Dr. Norwood placed NASA's SBIR program in the context of the agency's overall involvement with small business. Such activities come to about $1 billion annually, with between $500 million and $700 million going to firms in the high-technology sector.

Kesh Narayanan
National Science Foundation

Promising to enumerate the strengths, weaknesses, and opportunities associated with NSF's SBIR efforts, Dr. Narayanan began by describing the program's external advisory board and naming three conference attendees who serve on it: Terry Bibbens and Dan Hill of the Small Business Administration and Ann Eskesen of the Innovative Development Institute. This body looks at the NSF program in a very critical manner—in the positive as well as the negative sense—and program officials not only pay attention to its advice but implement its recommendations.

Proposal Selection

Dr. Narayanan said that SBIR relies on the same method for proposal selection that is applied in all programs at NSF: external peer review. The fact that not only program officers but members of the outside research community evaluate the proposals helps ensure that proposals selected are of high quality. And the outside reviewers, a very large percentage of whom are academics, have uniformly praised the quality of the proposals submitted. He observed that this is "at odds with the general noise that you may hear about this gap between the academic community and small business." If there is a dichotomy, he said, it is between science and engineering in the composition of the proposals received: Although only 10 percent of NSF's activities are in engineering, that field's portion of NSF's SBIR program is about 40 percent.

Weaknesses of SBIR

Dr. Narayanan acknowledged the perception within NSF that SBIR funding amounts to a tax on R&D resources. A second weakness is the low volume of proposals received by its STTR program, which mandates a partnership. And although university involvement in NSF's SBIR program is, at 40 percent of proposals, fairly high, there is room for improvement in this case as well: Much of the academic participation takes the form of faculty consulting, and actual dollars invested remains low.

Dr. Narayanan attempted to build the NSF and academic communities' involvement in the agency's SBIR program by reminding reviewers that, as the ones who selected the projects, they share in their successes. Making the SBIR program subject to the Government Performance and Results Act offers another opportunity for building inclusiveness among the academic, NSF, and small business communities. Finally, as Rita Colwell, nominated to succeed Neal Lane as director of NSF, has had extensive experience with SBIR, he foresees significant gains for the agency's program.

John Williams
U.S. Navy

The Need for Flexibility in SBIR

Mr. Williams began by warning against the imposition of government-wide rules that would standardize SBIR, arguing that the program's flexibility allows it to respond to the specific agencies' needs. He stressed that, because SBIR funds are derived from a "tax" on an agency's R&D budget, their disbursement should be geared toward benefiting that agency's research program directly. But this principle may become obscured: Whereas the emphasis of the Navy's SBIR program is on development of products the service itself will buy rather than those whose primary market is the private sector, the term "commercialization" is often taken to refer mainly to the latter—even, at times by the Navy's own evaluators. "We need to do a better job of training people," Mr. Williams admitted, but he nonetheless argued that agencies in general should keep in mind "where the money is coming from" when deciding how to spend it.

Measures of Success

Similarly, the program's success will be measured differently by different agencies. Setting a specific time within which products must be commercialized as a measure of SBIR success would be a mistake, Mr. Williams suggested. A Defense Department agency, for example, may need to get a program into its budget before it is able to buy a technology, something which may not be true across the federal government. He also reminded the audience that, when R&D proves that an idea is not technically feasible, this can be valid R&D. But it may be "viewed as a negative" in evaluating the success of a program. He also urged caution in obliging program managers to spend a great deal of time measuring the impact of their activities, saying that this would impinge on their ability to work with companies and technical monitors. This may loom even larger as R&D

budgets decline, leaving "less and less money [that would allow] program managers from the government to work with the technical officers of the companies."

Returning to commercialization, Mr. Williams observed that many who evaluate SBIR proposals for the government agencies are far more qualified to judge technologies than business plans. Of the companies he has reviewed, "many are very strong technically but have not thought out how they are going to do their marketing plans." He suggested that pointing firms toward economic development groups in their areas might help them acquire better tools for crafting marketing plans. But he warned that the problem of finding a government evaluator skilled in reviewing such material might remain.

DISCUSSION

Working with the States

Brian Belanger of the Advanced Technology Program at the National Institute of Standards and Technology observed that state economic development agencies now have "very aggressive" programs to help high-technology firms with commercialization. He asked whether the state agencies' role in working with companies is growing. Dan Hill of the Small Business Administration responded by noting that a group he co-chairs under the U.S. Innovation Partnership—a joint effort of the White House, the National Governors Association, and the U.S. Department of Commerce—is looking into how the SBIR program might be leveraged with state and local resources, including the public partnerships that have been formed in many states.

"My personal view is that we in Washington tend to think we have all the answers," said Mr. Hill, who also serves, along with Dr. Davis, on the National Science and Technology Council's Subcommittee on Innovation Partnerships. "In fact, we really do not. What we need to do is to get better at listening to the states, learning what's hot in the technology fields, getting ideas from them on what they are supporting, and then working together to leverage those dollars. And that's what we are trying to do in our groups."

Agency-Awardee Interactions

Mark Crawford, a reporter with *New Technology Week*, asked Dr. Norwood, who had mentioned not having enough funding to travel to SBIR applicant and awardee firms, whether he had the latitude to require them to visit him at NASA headquarters. Dr. Norwood stated that because the administration of SBIR must be funded out of other R&D management accounts, such spending takes away from other activities, a fact that imposes limits on it. Second, as NASA's many SBIR Phase II awardees are spread across the country, the problem of keeping in touch with them is in itself daunting. He has observed that venture-capital companies tend to fund a "well-defined set" of firms that is generally limited to a particular geographic "sphere in which they have some influence—perhaps a technology region like Silicon Valley." In this way, they are able to have access to their companies on a daily basis; he questioned, however, whether this approach offers a practicable solution for an agency with a widespread constituency.

Responding to the same question, Ms. de Blanc noted that the DoE's travel budget is included in its personnel budget, which has shrunk with the extensive reductions in force of the past two years. "We have to prioritize travel funds with exquisite discretion," she said. "It gets down to what are you going to do—go out and review an entire national laboratory or four SBIR projects" of the 350 that the department is funding? Increasing the size of individual SBIR grants is under consideration, which in some areas would make it more reasonable to devote funds to personal oversight of the projects. Nevertheless, DoE prefers that grant recipients' use their funding for research rather than spending it on trips to meet program officials.

Conclusion

Mark Myers
Xerox Corporation

Dr. Myers began his closing observations by saying that, as a member of the STEP Board and an official of a large corporation, he had not approached the symposium with a strong point of view. However, he said the day contained "a very interesting and engaging set of conversations."

He inferred from SBIR's strong support in both the Congress and the administration that the program "plays well both inside and outside the Beltway." This broad-based support is earned, given SBIR's history of supporting not only the growth of jobs and the overall economy, but also the missions of participating agencies. Agency missions have been furthered by specific results of the program's R&D activities. He praised the SBIR's focus on technical merit and its consequent incorporation of competition and peer review, which he called "time-proven approaches to assure quality."

The Need for More Research and Evaluation

Nevertheless, the SBIR program's very significance underscores the need for more research on how it is working and how it should work. Dr. Myers looked to the dialogue that would follow from the symposium for amplification of these issues, and he emphasized the commitment of those in attendance, suggesting that it would not be hard to find less committed constituencies for programs that enjoy a similar level of funding. Pointing out that the definition of "commercialization" has been ambiguous within the context of SBIR, he pointed to the need for clarification in view of the fact that it "seems to be a place where people stumble." Similarly, the participating agencies' autonomy in managing their versions of the SBIR program means that "on a de facto basis, there are 'multiple programs'" pursuing a variety of goals. Although the argument for diversity is strong, the program will be considered as a whole in the upcoming reauthorization and it must therefore make sense as a whole.

The Need for Fast Decision-Making

From the point of view of the commercial sector, Dr. Myers advised the audience, "everything you can do to speed up anything, the better off you are going to be." He acknowledged that the issue of timing could be a particularly problematic one in some areas of energy and in regulated industries such as health care. But he recalled that Hewlett-Packard, a $37 billion company, derives over 60 percent of its revenues from products that have been in the marketplace for only 18 months. Pointing out that the speed of product introduction is a basis of competition, he said speed defines what makes a technology "commercial," at least in the information technology sector.

The Need for Small Business Capital

Stressing the need for variety in funding sources for small business, Dr. Myers noted the SBIR program's role as an "alternative" funding source. He observed that SBIR funding is "patient" in that it does not have to be renewed annually, which has the value of enabling a sustained approach, at least for individual awardees. As for the level of SBIR resources, he recommended turning from the matter of what percentage of federal R&D funding the program should receive to the question: "Can the magnitude of $1.2 billion meaningfully grow to $2 billion?" Such an increase would raise further issues regarding management, assessment, and validation processes.

The perception on the participating agencies' part that SBIR funding is a tax on their R&D budgets and competes with the interests arising directly from their missions will have to be addressed in any discussion of increasing the program's resources, or even of simply extending them through reauthorization. Another question that requires attention is the accountability of SBIR companies, Dr. Myers said. In particular, he recommended consideration be given to limiting awards to firms that had won previously but failed to commercialize.

Speaking on behalf of the STEP, Dr. Myers placed the SBIR program firmly within the broad area of government-industry partnership projects that the Board is engaged in evaluating. He expressed the Board's appreciation to the attendees for their contributions to the day's discussion.

IV

ANNEXES

Annex A

"Public Venture Capital": Rationales and Evaluation

*Joshua Lerner**

Within the past few years, public efforts to finance small high-technology firms have proliferated. This article reviews the motivations for and the assessment of these efforts. It explores the underlying challenges that the financing of young growth firms poses, the ways that specialized financial intermediaries address them, and the rationales for public efforts to finance these companies. The final section raises a set of questions about the assessment of these efforts.

INTRODUCTION

The federal government has played an active role in financing new firms, particularly in high-technology industries, since the Soviet Union's launch of the Sputnik satellite. In recent years, European and Asian nations and many U.S. states have adopted similar initiatives. While these programs' precise structures have differed, the efforts have been predicated on two shared assumptions: *(i)* that the private sector provides insufficient capital to new firms, and *(ii)* that the government either can identify investments which will ultimately yield high so-

*Harvard University and National Bureau of Economic Research. This is based on comments at the "Symposium on the SBIR Program" at the National Academy of Sciences, as well as on conversations with Zoltan Acs, Ken Flamm, Paul Gompers, Adam Jaffe, Bill Sahlman, Greg Udell, and Chuck Wessner. Parts of this article are adapted from Lerner [1996] and Lerner [1998]. Financial support was provided by Harvard Business School's Division of Research. All errors are my own.

cial and/or private returns or can encourage effective financial intermediaries.[1] In contrast to many forms of government intervention designed to boost economic growth, such as privatization programs, these claims have received little scrutiny by economists.

The neglect of these questions is unfortunate. While the sums of money involved are modest relative to public expenditures on defense procurement or retiree benefits, these programs are very substantial when compared to contemporaneous private investments in new firms. Several examples, documented in Gompers and Lerner [1998b], underscore this point:

- the Small Business Investment Company (SBIC) program led to the provision of more than $3 billion to young firms between 1958 and 1969, more than three times the total private venture capital investment during these years (Noone and Rubel [1970]).
- in 1995, the sum of the equity financing provided through and guaranteed by public small business financing programs was $2.4 billion, more than 60% of the amount disbursed by traditional venture funds in that year. Perhaps more significantly, the bulk of the public funds went to early-stage firms, which in the past decade had accounted for only about 30% of the disbursements by independent venture capital funds (Venture Economics [1996]).
- some of America's most dynamic technology companies received support through the SBIC and Small Business Innovation Research (SBIR) programs while still privately held entities, including Apple Computer, Chiron, Compaq, and Intel.
- public venture capital programs have also had a significant impact overseas: e.g., Germany has created about 800 federal and state government financing programs for new firms over the past two decades, which provide the bulk of the financing for technology-intensive start-ups (Organization for Economic Cooperation and Development [1995]).

Government programs in this arena have been divided between those efforts that directly fund entrepreneurial firms and those that encourage or subsidize the development of outside investors.

While these efforts have proliferated, a consensus as to how to evaluate these programs remains elusive. The gap between the approaches employed by academics and practitioners is substantial. Furthermore, there is a lack of consensus among economists as to what the proper approaches are.

[1] It is striking to note the similar emphasis on these rationales in, for instance, the statement of Senator John Sparkman [1958] upon the passage of the Small Business Investment Act and the recent testimony of Dr. Mary Good, Under Secretary for Technology at the U.S. Department of Commerce [1995]. The rationales for such programs are discussed in depth in U.S. Congressional Budget Office [1985].

This article will provide an overview of the motivations for these efforts to encourage individual investors. In Section 2, the underlying challenges that the financing of young growth firms poses are discussed, as well as the ways that specialized financial intermediaries address them. The rationales for public programs are explored in Section 3. Section 4 concludes the paper and raises a set of questions about the assessment of these efforts.

VENTURE CAPITALISTS AND THE FINANCING CHALLENGE

The initial reaction of a financial economist to the argument that the government needs to invest in young firms is likely to be skepticism. A lengthy literature has highlighted the role of financial intermediaries in alleviating moral hazard and information asymmetries.[2] Young high-technology firms are often characterized by considerable uncertainty and informational asymmetries, which permit opportunistic behavior by entrepreneurs. Why one would want to encourage public officials instead of specialized financial intermediaries (venture capital organizations) as a source of capital in this setting is not immediately obvious.

The Challenge of Financing Young High-Technology Firms

To briefly review the types of conflicts that can emerge in these settings, Jensen and Meckling [1976] demonstrate that agency conflicts between managers and investors can affect the willingness of both debt and equity holders to provide capital. If the firm raises equity from outside investors, the manager has an incentive to engage in wasteful expenditures (e.g., lavish offices) because he does not bear their entire cost. Similarly, if the firm raises debt, the manager may increase risk to undesirable levels. Because providers of capital recognize these problems, outside investors demand a higher rate of return than would be the case if the funds were internally generated.

Even if the manager is motivated to maximize shareholder value, informational asymmetries may make raising external capital more expensive or even preclude it entirely. Myers and Majluf [1984] and Greenwald, Stiglitz, and Weiss [1984] demonstrate that equity offerings of firms may be associated with a "lemons" problem (first identified by Akerlof [1970]). If the manager is better informed about the investment opportunities of their firms than the investors and

[2] Editors Note: Moral hazard refers to inefficient behavior by one actor in a transaction brought on by differences in information available to parties in the transaction. A classic example is insurance. An individual who purchases insurance has an incentive to engage in risky behavior, because insurance will compensate for negative consequences of the behavior. The insurance company might address the problem by collecting information and closely monitoring the individual. However, it is likely to be costly, and in some cases impossible, to collect sufficient information to adequately monitor the individual's behavior. See Dennis Carlton and Jeffrey Perloff, *Industrial Organization,* New York: Prentice, 1992, pp. 716-717.

acts in the interest of current shareholders, then the manager issues new shares only when the company's stock is overvalued. Indeed, numerous studies have documented that stock prices decline upon the announcement of equity issues, largely because of the negative signal sent to the market.

These information problems have also been shown to exist in debt markets. Stiglitz and Weiss [1981] show that if banks find it difficult to discriminate among companies, raising interest rates can have perverse selection effects. In particular, the high interest rates discourage all but the highest-risk borrowers, so the quality of the loan pool declines markedly. To address this problem, banks may restrict the amount of lending rather than increasing interest rates.

These problems in the debt and equity markets are a consequence of the information gaps between the entrepreneurs and investors. If the information asymmetries could be eliminated, financing constraints would disappear. Financial economists argue that specialized financial intermediaries can address these problems. By intensively scrutinizing firms before providing capital and then monitoring them afterwards, they can alleviate some of the information gaps and reduce capital constraints.

Responses by Venture Capitalists

The financial intermediary that specializes in funding young high-technology firms is the venture capital organization. The first modern venture capital firm, American Research and Development (ARD), was formed in 1946 by MIT President Karl Compton, Harvard Business School Professor Georges F. Doriot, and local business leaders. A small group of venture capitalists made high-risk investments in emerging companies that were formed to commercialize technology developed for World War II. The success of the investments ranged widely: almost half of ARD's profits during its 26-year existence as an independent entity came from its $70,000 investment in Digital Equipment Company (DEC) in 1957, which grew in value to $355 million. Because institutional investors were reluctant to invest, ARD was structured as a publicly traded closed-end fund and marketed mostly to individuals (Liles [1977]). The few other venture organizations begun in the decade after ARD's formation were also structured as closed-end funds.

The first venture capital limited partnership, Draper, Gaither, and Anderson, was formed in 1958. Imitators soon followed, but limited partnerships accounted for a minority of the venture pool during the 1960s and 1970s. Most venture organizations raised money either through closed-end funds or small business investment companies (SBICs), federally guaranteed risk capital pools that proliferated during the 1960s. While investor demand for SBICs in the late 1960s and early 1970s was strong, incentive problems ultimately led to the collapse of the sector. The annual flow of money into venture capital during its first three decades never exceeded a few hundred million dollars and usually was substantially less.

The activity in the venture industry increased dramatically in late 1970s and early 1980s. Industry observers attributed much of the shift to the U.S. Department of Labor's clarification of ERISA's "prudent man" rule in 1979. Prior to that year, the Employee Retirement Income Security Act (ERISA) limited pension funds from investing substantial amounts of money in venture capital or other high-risk asset classes. The Department of Labor's clarification of the rule explicitly allowed pension managers to invest in high-risk assets, including venture capital. In 1978, when $424 million was invested in new venture capital funds, individuals accounted for the largest share (32 percent). Pension funds supplied just 15 percent. Eight years later, when more than $4 billion was raised, pension funds accounted for more than half of all contributions. (These annual commitments represent pledges of capital to venture funds raised in a given year. This money is typically invested over three to five years starting in the year the fund is formed.)

The subsequent years saw both very good and trying times for venture capitalists. On the one hand, venture capitalists have backed during the 1980s and 1990s many of the most successful high-technology companies, including Apple Computer, Cisco Systems, Genentech, Netscape, and Sun Microsystems. A substantial number of service firms (including Staples, Starbucks, and TCBY) have also received venture financing. At the same time, commitments to the venture capital industry were very uneven. The annual flow of money into venture funds increased by a factor of ten during the early 1980s, peaking at just under six billion 1996 dollars. From 1987 through 1991, however, fund-raising steadily declined. Over the past five years, the pattern has been reversed; 1997 represented a record fund-raising year, in which nearly $10 billion was raised by venture capitalists. This process of rapid growth and decline has created a great deal of instability in the industry.

To address the information problems that preclude other investors in small high-technology firms, the partners at venture capital organizations employ a variety of mechanisms. First, business plans are intensively scrutinized: of those firms that submit business plans to venture capital organizations, historically only 1% has been funded (Fenn, Liang, and Prowse [1995]). The decision to invest is frequently made conditional on the identification of a syndication partner who agrees that this is an attractive investment (Lerner [1994]). In exchange for their capital, the venture capital investors demand preferred stock with numerous restrictive covenants and representation on the board of directors.

Once the decision to invest is made, the venture capitalists frequently disburse funds in stages. Managers of these venture-backed firms are forced to return repeatedly to their financiers for additional capital in order to ensure that the money is not squandered on unprofitable projects. In addition, venture capitalists intensively monitor managers, often contacting firms on a daily basis and holding monthly board meetings during which extensive reviews of every aspect of the firm are conducted. (Various aspects of the oversight role played by venture capitalists are documented in Gompers [1995], Lerner [1995], and Sahlman [1990]; the theoretical literature is reviewed in Barry [1994].)

It is important to note that, even with these many mechanisms, the most likely primary outcome of a venture-backed investment is failure, or at best modest success. Gompers [1995] documents that out of a sample of 794 venture capital investments made over three decades, only 22.5% ultimately succeeded in going public, the avenue through which venture capitalists typically exit their successful investments. (A Venture Economics study [1988] finds that a $1 investment in a firm that goes public provides an average cash return to venture capitalists of $1.95 in excess of the initial investment, with an average holding period of 4.2 years. The next best alternative, a similar investment in an acquired firm, yields a cash return of only 40 cents over a 3.7-year mean holding period.) Similar results emerge from Huntsman and Hoban's [1980] analysis of the returns from 110 investments by three venture capital organizations. About one in six investments was a complete loss, while 45% were either losses or simply broke even. The elimination of the top-performing 9% of the investments was sufficient to turn a 19% gross rate of return into a negative return.

In short, the environment in which venture organizations operate is extremely difficult. It is the mechanisms that are bundled with the venture capitalists' funds that are critical in assuring that they receive a satisfactory return. These circumstances have led to venture capital organizations emerging as the dominant form of equity financing for privately held technology-intensive businesses.[3]

RATIONALES FOR PUBLIC PROGRAMS

At the same time, there are reasons to believe that despite the presence of venture capital funds, there still might be a role for public venture capital programs. In this section, I assess these claims. I highlight two arguments: that public venture capital programs may play an important role by certifying firms to outside investors, and that these programs may encourage technological spillovers.

The Certification Hypothesis

A growing body of empirical research suggests that new firms, especially technology-intensive ones, may receive insufficient capital due to the informa-

[3] While evidence regarding the financing of these firms is imprecise, Freear and Wetzel's [1990] survey suggests that venture capital accounts for about two-thirds of the external equity financing raised by privately held technology-intensive businesses from private-sector sources.

tion problems discussed in the previous section.[4] If public venture capital awards could certify that firms are of high quality, these information problems could be overcome and investors could confidently invest in these firms.

As discussed above, venture capitalists specialize in financing these types of firms. They address these information problems through a variety of mechanisms. Many of the studies that document capital-raising problems examine firms during the 1970s and early 1980s, when the venture capital pool was relatively modest in size. Since the pool of venture capital funds has grown dramatically in recent years (Gompers and Lerner [1996, 1998b]), even if small high-technology firms had numerous value-creating projects that they could not finance in the past, one might argue that it is not clear this problem remains today.

A response to this argument emphasizes the limitations of the venture capital industry. Venture capitalists back only a tiny fraction of the technology-oriented businesses begun each year. In 1996, a record year for venture disbursements, 628 companies received venture financing for the first time (VentureOne [1997]); to put this in perspective, the Small Business Administration estimates that in recent years close to one million businesses have been started annually. Furthermore, these funds have been very concentrated: 49% of venture funding in 1996 went to companies based in either California or Massachusetts, and 82% went to firms specializing in information technology and the life sciences (VentureOne [1997]).

It is not clear, however, what lessons to draw from these funding patterns. Concentrating investments in such a manner may well be an appropriate response to the nature of opportunities. Consider, for instance, the geographic concentration of awards. Recent models of economic growth—building on earlier works by economic geographers—have emphasized powerful reasons why successful high-technology firms may be very concentrated. The literature highlights several factors that lead similar firms to cluster in particular regions, including knowledge spillovers, specialized labor markets, and the presence of critical intermediate goods producers.[5] Case studies of the development of high-technology regions (e.g., Saxenian [1994]) have emphasized the importance of such intermediaries as venture capitalists, lawyers, and accountants in facilitating this clustering.

A related argument for public investments is that the structure of venture investments may make them inappropriate for many young firms. Venture funds tend to make quite substantial investments, even in young firms; the mean venture investment in a start-up or early-stage business between 1961 and 1992 (expressed in 1996 dollars) was $2.0 million (Gompers [1995]). The substantial

[4] The literature on capital constraints (reviewed in Hubbard [1998]) documents that an inability to obtain external financing limits many forms of business investment. Particularly relevant are works by Hall [1992], Hao and Jaffe [1993], and Himmelberg and Petersen [1994]. These show that capital constraints appear to limit research-and-development expenditures, especially in smaller firms.

[5] The theoretical rationales for such effects are summarized in Krugman [1991].

size of these investments may be partially a consequence of the demands of institutional investors. The typical venture organization raises a fund (structured as a limited partnership) every few years. Because investments in partnerships are often time-consuming to negotiate and monitor, institutions (limited partners) prefer making relatively large investments in venture funds, typically $10 million or more. Furthermore, governance and regulatory considerations lead institutions to limit the share of any fund that any one limited partner holds.[6] As a consequence, venture organizations typically raise substantial funds of $100 million or more. Because each firm in his portfolio must be closely scrutinized, the typical venture capitalist is typically responsible for no more than a dozen investments. Venture organizations are consequently unwilling to invest in very young firms that require only small capital infusions.[7]

This problem may be increasing in severity with the growth of the venture industry, as discussed above. As the number of dollars per venture fund and dollars per venture partner have grown, so too has the size of venture investments. For instance, the mean financing round for a start-up firm has climbed (in 1996 dollars) from $1.6 million in 1991 to $3.2 million in 1996 (VentureOne [1997]).

Again, it is not clear what lessons to draw from these financing patterns. Venture capitalists may have eschewed small investments because they were simply not profitable, because of either the high costs associated with these transactions or the poor prospects of the thinly capitalized firms.[8] Encouraging public investments in small firms may be counter-productive and socially wasteful if the financial returns are unsatisfactory and the companies financed are not viable.

Support for these claims is found in recent work on the long-run performance of initial public offerings (IPOs). Brav and Gompers [1997] show that IPOs that had previously received equity financing from venture capitalists outperform other offerings. These findings underscore concerns about policies which seek to encourage public investments in companies that are rejected by professional investors.

Furthermore, it appears that there were in 1997 a number of financial innovations to address the needs of early-stage entrepreneurs. These included the

[6] The structure of venture partnerships is discussed at length in Gompers and Lerner [1996, 1998].

[7] There are two primary reasons that venture funds do not simply hire more partners if they raise additional capital. First, the supply of venture capitalists is quite inelastic. The effective oversight of young companies requires highly specialized skills that can only be developed with years of experience. A second important factor is the economics of venture partnerships. The typical venture fund receives a substantial share of its compensation from the annual fee, which is typically between 2% and 3% of the capital under management. This motivates venture organizations to increase the capital that each partner manages.

[8] For a theoretical discussion of why poorly capitalized firms are less likely to be successful, see Bolton and Scharfstein [1990].

creation of incubators and "entrepreneur-in-residence" programs by established venture organizations such as Mayfield and Mohr Davidow. Other examples are innovative efforts to direct the resources of individual investors to small venture capital funds (an example is Next Generation Partners, a "fund-of-funds" for wealthy families developed by FLAG Venture Partners). Finally, some institutional investors are displaying an increased willingness to provide capital to first time and seed venture funds. Thus, market forces may be addressing whatever problem has existed.

The Presence of R&D Spillovers

A second rationale emerges from the literature on R&D spillovers. Public finance theory emphasizes that subsidies are an appropriate response in the case of activities that generate positive externalities. Such investments as R&D expenditures and pollution control equipment purchases may have positive spillovers that help other firms or society as a whole. Because the firms making the investments are unlikely to capture all the benefits, public subsidies may be appropriate.

An extensive literature (reviewed in Griliches [1992] and Jaffe [1996]) has documented the presence of R&D spillovers. These spillovers take several forms. For instance, the rents associated with innovations may accrue to competitors who rapidly introduce imitations, developers of complementary products, or to the consumers of these products. Whatever the mechanism of the spillover, however, the consequence is the same: the firm invests below the social optimum in R&D.

After reviewing a wide variety of studies, Griliches estimates that the gap between the private and social rate of return is substantial: the gap is probably equal to between 50% and 100% of the private rate of return. While few studies have examined how these gaps vary with firm characteristics, a number of case-based analyses (Jewkes et al. [1958], Mansfield et al. [1977]) suggest that spillover problems are particularly severe among small firms. These organizations may be particularly unlikely to effectively defend their intellectual property positions or to extract most of the rents in the product market.

Even if these problems are substantial, however, the government may not be able to address them dispassionately. An extensive political economy and public finance literature has emphasized the possible distortion that may result from government subsidies as particular interest groups or politicians seek to direct subsidies in a manner that benefits themselves. As articulated by Olson [1965] and Stigler [1971], and formally modeled in works such as Peltzman [1976] and Becker [1983], the theory of regulatory capture suggests that direct and indirect subsidies will be captured by parties whose joint political activity is not too difficult to arrange (i.e., when "free-riding" by coalition members is not too large a problem).

These distortions may manifest themselves in several ways. One possibility (discussed, for instance, in Eisinger [1988]), is that firms may seek transfer payments that directly increase their profits. Politicians may acquiesce in such transfers in the case of companies that are politically connected. A more subtle distortion is discussed by Cohen and Noll [1991] and Wallsten [1996]: officials may seek to select firms based on their likely success, and fund them regardless of whether the government funds are needed. In this case, they can claim credit for the firms' ultimate success even if the marginal contribution of the public funds was very low.

THE CHALLENGE OF EVALUATION

As public venture capital programs have increased in number, policymakers and economists are increasingly grappling with the question of how to assess these programs. Not only do substantial divisions exist between the approaches employed by academics and practitioners, but there is little consensus within the academic community itself about the best evaluation methodologies. In this final section, I will review some of the most frequently encountered approaches and discuss their strengths and limitations. I will discuss their implementation in the context of the program that is the subject of this symposium, the Small Business Innovation Research (SBIR) program.

The approaches most frequently employed by practitioners have the virtue of being relatively straightforward to implement and communicate. One approach—utilized by many agencies when examining their SBIR programs—has been to highlight successful firms.[9] Another approach has been to survey firms that have been funded under the SBIR program, asking such questions as whether the technologies funded were ever commercialized, the extent to which their development would have occurred without the public award, and how firms assessed their experiences with the program more generally.[10]

These approaches have important limitations. First, many awardees may have a stake in the programs that have funded them, and consequently feel inclined to give favorable answers (i.e., that they have received benefits from the program and that commercialization would not have taken place without the awards). This may be a particular problem in the case of the SBIR initiative, since many small high-technology company executives have organized to lobby for its renewal. Second, in other cases, the results may be biased the other way: firms may be unwilling to acknowledge that they received important benefits

[9] In the context of the SBIR program, see U.S. Small Business Administration [1994], and many agency publications.

[10] Examples of evaluations of the SBIR program include Myers, Stern, and Rorke [1983], Price, Waterhouse [1985], and U.S. General Accounting Office [1987, 1989, 1992].

from participating in public programs, lest they attract unwelcome attention. This is especially likely to be a problem in the life sciences, since periodic press and Congressional investigations have highlighted "give-aways" of research funded by the National Institutes of Health to biotechnology and pharmaceutical companies. Third, in many cases, it may simply be very difficult to identify the marginal contribution of a public venture capital award, which may be one of many sources of financing that a firm employed to develop a given technology. Finally, as argued by Wallsten [1996], these evaluation criteria may have a distorting effect on which firms are selected for participation in these programs, leading to an emphasis on "safe" firms that would have succeeded anyway.

The approaches employed by academics have important limitations as well. The most common approach is to examine in a regression framework the marginal impact of public funding on private research spending. Studies of federal technology programs by academic economists, beginning with Levy and Terleckyj [1983], have tended to focus on the short-run effects of these efforts. In particular, they often ask whether federal funds substitute for or stimulate private R&D spending. In another application, Irwin and Klenow [1996] show that semiconductor manufacturers substantially reduced their own R&D spending while participating in the Sematech consortium. In theory, these frameworks should be applicable to the assessment of public venture capital programs. Indeed, Wallsten [1996] shows that the subset of SBIR awardees that were publicly traded reduced their own spending on R&D in the years immediately following the award.

However valuable a framework it may be when examining the macroeconomic impact of public expenditures, it is less clear that this econometric approach is appropriate when assessing public efforts to assist small high-technology firms. In many cases, small high-technology firms are organized around one key scientist or engineer and his research laboratory or product development team. It may not be possible to accelerate the project's progress by "scaling up" the project through the addition of additional researchers or technicians. It may well be rational for a firm not to increase its rate of spending, but rather to use the funds to prolong the time before it needs to seek additional capital. To interpret such a short-run reduction in other research spending as a negative signal is very problematic.

A second academic approach is to examine the long-run impact of participation in public venture capital programs on the growth of the firms themselves, relative to a matched set of firms. In this way, it is possible to assess whether either superior firms were selected for the program or participation in the program was associated with ultimate success, although disentangling the two effects, as discussed below, is challenging. In the context of the SBIR program, Lerner [1996] analyzes the growth of 1435 SBIR awardees and matching firms over a ten-year period and documents that the awardees appear to have superior employment growth.

This approach also has some important limitations. Most fundamentally, policymakers should seek to maximize social, not private, returns. If the growth of the SBIR awardees is merely at the expense of their rivals, the impact of the program on public welfare is likely to be minimal. Second, even the measures of private benefits that can be employed are imperfect. Ideally, the increase in firm value would be measured. Unfortunately, over 98% of the firms were privately held at the time of their first SBIR award. Consequently, assessing the valuation and profitability of these awards is very difficult. Thus, Lerner's examination is confined to two measures that are only imperfectly corrrelated with firm value, employment and sales. Finally, it is difficult to disentangle whether the superior performance of the awardees is due to the selection of better firms or the positive impact of the awards.[11]

Thus, the evaluation of public venture capital programs remains a subject of lively debate. The choice of appropriate valuation methodologies is likely to be of considerable interest for both academics and practitioners in the years to come. Efforts to encourage discussions of these issues between academics and practitioners, such as the ongoing joint venture between the U.S. Department of Commerce's Advanced Technology Program and the National Bureau of Economic Research, should be encouraged. This type of healthy dialog should lead to more effective programs.

REFERENCES

Akerlof, G.A., 1970, The market for 'lemons': Qualitative uncertainty and the market mechanism, Quarterly Journal of Economics 84, 488-500.

Barry, C.B., 1994, New directions in research on venture capital finance, Financial Management 23 (Autumn), 3-15.

Becker, G.S., 1983, A theory of competition among pressure groups for political influence, Quarterly Journal of Economics 98, 371-400.

[11]Lerner [1996] tries to address this issue in a supplemental analysis using the following argument: firms whose key assets are intangible intellectual property are much harder for outside investors to evaluate using traditional financial measures. If SBIR awards are certifying firm quality to outside investors, then these signals may be particularly valuable in these industries. SBIR awards should then be more strongly associated with firm growth in high-technology industries. An alternative hypothesis is that Federal officials are selecting firms likely to grow rapidly, even without public subsidies. A potential motive would be that politicians could claim credit for the firms' ultimate success, even if the marginal contribution of the public funds was very low. Though the insights of Federal officials may give them a greater insight relative to that of other investors (and thus make a signal more valuable), it is by no means certain that it is easier to select successful firms in these industries. Empirical studies suggest that predicting success is much more difficult in high-technology industries. This suggests the reverse pattern: SBIR awards should be more correlated with firm growth in low-technology industries. Consistent with the certification hypothesis, he finds that the relationship between SBIR awards and growth is much stronger in high-technology industries.

Bolton, P., and D. Scharfstein, 1990, A theory of predation based on agency problems in financial contracting, American Economic Review 80, 93-106.

Brav, A., and P.A. Gompers, 1997, Myth or reality?: Long-run underperformance of initial public offerings; Evidence from venture capital and nonventure capital-backed IPOs, Journal of Finance, 52, 1791-1821.

Cohen, L.R., and R.G. Noll, editors, 1991, The technology pork barrel (Washington, Brookings Institution).

Eisinger, P.K., 1988, The rise of the entrepreneurial state: State and local economic development policy in the United States (Madison, University of Wisconsin Press).

Fenn, G.W., N. Liang, and S. Prowse, 1995, The economics of the private equity market (Washington, Board of Governors of the Federal Reserve System).

Freear, J., and W.E. Wetzel, Jr., 1990, Who bankrolls high-tech entrepreneurs?, Journal of Business Venturing 5, 77-89.

Gompers, P.A., 1995, Optimal investment, monitoring, and the staging of venture capital, Journal of Finance 50, 1461-1489.

Gompers, P.A., and Lerner, J., 1996, The use of covenants: An empirical analysis of venture partnership agreements. Journal of Law and Economics 39, 463-498.

__________, and __________, 1998a, An analysis of compensation in the U.S. venture capital partnership, Journal of Financial Economics, forthcoming.

__________, and __________, 1998b, What drives venture capital fund-raising? (Unpublished working paper, Harvard University).

Good, M.L., 1995, Prepared testimony before the Senate Commerce, Science and Transportation Committee, Subcommittee on Science, Technology and Space (Photocopy, U.S. Department of Commerce).

Greenwald, B.C., J.E. Stiglitz, and A. Weiss, 1984, Information imperfections in the capital market and macroeconomic fluctuations, American Economic Review Papers and Proceedings 74, 194-199.

Griliches, Z., 1992, The search for R&D spillovers, Scandinavian Journal of Economics 94, S29-S47.

Hall, B.H., 1992, Investment and research and development: Does the source of financing matter? (Working Paper No. 92-194, Department of Economics, University of California at Berkeley).

Hao, K.Y., and A.B. Jaffe, 1993, Effect of liquidity on firms' R&D spending, Economics of Innovation and New Technology 2, 275-282.

Himmelberg, C.P., and B.C. Petersen, 1994, R&D and internal finance: A panel study of small firms in high-tech industries, Review of Economics and Statistics 76, 38-51.

Hubbard, R.G., 1998, Capital-market imperfections and investment, Journal of Economic Literature, 36, 193-225.

Huntsman, B., and J.P. Hoban, Jr., 1980, Investment in new enterprise: Some empirical observations on risk, return, and market structure, Financial Management 9 (Summer) 44-51.

Irwin, D.A., and P.J. Klenow, 1996, High tech R&D subsidies: Estimating the effects of Sematech, Journal of International Economics 40, 323-344.

Jaffe, A.B., 1996, Economic analysis of research spillovers—Implications for the Advanced Technology Program (Washington, Advanced Technology Program, National Institute of Standards and Technology, U.S. Department of Commerce).

Jensen, M.C., and W.H. Meckling, 1976, Theory of the firm: Managerial behavior, agency costs and ownership structure, Journal of Financial Economics 3, 305-360.

Jewkes, J., D. Sawers, and R. Stillerman, 1958, The sources of invention (New York, St. Martin's Press).

Krugman, P.R., 1991, Geography and trade (Cambridge, MIT Press).

Lerner, J., 1994, The syndication of venture capital investments, Financial Management 23 (Autumn) 16-27.

_________, 1995, Venture capital and the oversight of private firms, Journal of Finance 50, 301-318.

_________, 1996, The government as venture capitalist: The long-run effects of the SBIR program (Working paper no. 5753, National Bureau of Economic Research).

_________, 1998, Angel financing and public policy: An overview, *Journal of Banking and Finance*, 22 (August), 773-783.

Levy, D.M., and N. Terleckyj, 1983, Effects of government R&D on private R&D investment and productivity: A macroeconomic analysis, Bell Journal of Economics 14, 551-561.

Liles, P., 1977, Sustaining the venture capital firm (Cambridge, Management Analysis Center).

Mansfield, E., J. Rapoport, A. Romeo, S. Wagner, and G. Beardsley, 1977, Social and private rates of return from industrial innovations, Quarterly Journal of Economics 91, 221-240.

Myers, S.C., and N. Majluf, 1984, Corporate financing and investment decisions when firms have information that investors do not have, Journal of Financial Economics 13, 187-221.

Myers, S., R.L. Stern, and M.L. Rorke, 1983, A study of the Small Business Innovation Research program (Lake Forest, Illinois, Mohawk Research Corporation).

Noone, C.M., and S.M. Rubel, 1970, SBICs: Pioneers in organized venture capital (Chicago, Capital Publishing Company,).

Olson, M., 1965, The logic of collective action (Cambridge, Harvard University Press).

Organisation for Economic Co-operation and Development, 1995, Venture capital in OECD countries (Paris, Organisation for Economic Co-operation and Development).

Peltzman, S., 1976, Towards a more general theory of regulation, Journal of Law and Economics 19, 211-240.

Price, Waterhouse, 1985, Survey of small high-tech businesses shows Federal SBIR awards spurring job growth, commercial sales (Washington, Small Business High Technology Institute).

Sahlman, W.A., 1990, The structure and governance of venture capital organizations, Journal of Financial Economics 27, 473-521.

Saxenian, A., 1994, Regional advantage: Culture and competition in Silicon Valley and Route 128 (Cambridge, Harvard University Press).

Sparkman, J., 1958, Introduction, in: U.S. Congress, Senate, Small Business Committee, Small Business Investment Act of 1958 (Washington, U.S. Government Printing Office).

Stigler, G., 1971, The economic theory of regulation, Bell Journal of Economics 2, 3-21.

Stiglitz, J.E., and A. Weiss, 1981, Credit rationing in markets with incomplete information, American Economic Review 71, 393-409.

U.S. Congressional Budget Office, 1985, Federal financial support for high-technology industries (Washington, U.S. Congressional Budget Office).

U.S. General Accounting Office, 1987, Federal research: Small Business Innovation Research participants give program high marks (Washington, U.S. General Accounting Office).

_________, 1989, Federal research: Assessment of Small Business Innovation Research program (Washington, U.S. General Accounting Office).

_________, 1992, Federal research: Small Business Innovation Research shows success but can be strengthened (Washington, U.S. General Accounting Office).

U.S. Small Business Administration, 1994, Small Business Innovation Development Act: Tenth-year results (Washington, U.S. Government Printing Office) (and earlier years).

Venture Economics, 1988, Exiting venture capital investments (Wellesley, Venture Economics).

_________, 1996, Special report: Rose-colored asset class, Venture Capital Journal 36 (July) 32-34 (and earlier years).

VentureOne, 1997, National Venture Capital Association 1996 annual report (San Francisco, VentureOne).

Wallsten, S.J., 1996, The Small Business Innovation Research program: Encouraging technological innovation and commercialization in small firms? (Unpublished working paper, Stanford University).

Annex B

The Small Business Innovation Research Program and NSF SBIR Commercialization Results

*Roland Tibbetts**

Small high-technology firms are a large and growing national asset in terms of their innovative capacity and economic importance to the nation. Between 1985 and 1995, among small firms, the number of full-time equivalent R&D scientists and engineers in R&D performing companies increased 76 percent from 78,300 to 137,500. This compares with a 7 percent reduction during the same period in full-time equivalent R&D scientists and engineers from 437,500 to 408,500 in companies with 10,000 or more employees. Yet, in spite of excellent performance and growth in numbers, small firms have always had a difficult time obtaining R&D funding and receive only four percent of government R&D.

SBIR BACKGROUND

The forerunner of the Small Business Innovation Research Program (SBIR) was initiated at the National Science Foundation (NSF) in 1977 specifically to increase the opportunity of small high-technology companies to participate in NSF research and to convert research results into technological innovations and commercial applications for economic benefits to the nation. NSF was an excellent place to start because it funds advanced research in essentially all fields of science and engineering, except the medical and weapons areas. Early results at

* Mr. Tibbetts served as SBIR Program Manager at the National Science Foundation and played an instrumental role in developing the program. He submitted this paper as background material for the symposium. The views expressed in this paper are his own.

NSF were promising. As a result, in 1982 Congress passed legislation that required all 11 government departments and agencies with external R&D budgets of $100 million or more to initiate an SBIR program. The Small Business Administration coordinates this interagency effort. High-technology firms by definition must do high-risk research if they are to generate scientific and market breakthrough ideas that attract private investment. But this research often requires a degree of risk unacceptable to the private investor, including venture capitalists.

There was also a public need for a better mechanism to convert the results of federally funded R&D into commercial and economic benefits for the country. Prior to SBIR's creation, commercialization was not a high priority for government R&D managers. However, SBIR was intended to open up government R&D funding to the ideas and participation of innovative small firms. Its second major objective was to stimulate their conversion into commercial applications, by encouraging the use of private capital for the commercialization effort.

SBIR DESIGN

SBIR is a carefully designed three-phase, highly competitive interagency program for small firms that focuses specifically on government agency research needs and ideas that also may have commercial applications—and therefore economic benefits for the country.

Phase I is a short, six month project, currently for up to $100,000 (originally $50,000), depending on the agency. It supports advanced R&D to determine the technical feasibility of a research-based idea. It is a preliminary effort to determine, in the shortest period of time and with the least expenditure of government funds, whether the idea appears promising and whether the firm can do high quality research. Each SBIR agency describes its research needs in detail at least annually in comprehensive solicitations for proposals that also provide instructions to proposers. Proposals are reviewed by each agency's standard practice, such as by peer review at NSF, NIH, and some other agencies, or by in-house scientists and engineers, as in Defense and NASA.

Only projects with the most promising results in Phase I go on to a larger Phase II which is to advance the Phase I idea to a proof of concept prototype. It is often a two-year project for up to $750,000 (originally $500,000), depending on the agency. Phase II evaluation must take into consideration not only the scientific and technical merit of the project, but also its commercial potential. The latter is usually judged from a commercialization plan and written commitment for Phase III follow-on funding provided by the company, as well as the track record of the company in commercializing any previous SBIR projects it has received.

Both Phase I and Phase II cover all allowable costs and allow for a modest profit consistent with government R&D funding practices; both phases are subject to audit. SBIR policy provides world-wide commercial rights to resulting patents to the small firm. Government retains rights for its own use. Phase III is for product development to pursue potential commercial applications using private sector financing. Only if a government agency, such as the Department of Defense, is the potential customer for the Phase III results can government funding finance Phase III.

The risk of funding R&D ideas from small firms under SBIR is offset to some extent by the quality of the Federal technical review process and the sheer competitiveness of the program. On the average, NSF SBIR funds are available for only one out of every seven to eight Phase I proposals and only one out of three in Phase II. Only Phase I winners are eligible for Phase II. Only about five percent of proposals received at NSF in Phase I receive the larger Phase II award. These percentages have kept the quality of SBIR awards high, and this has contributed to respect for SBIR in both government and the private sector. Opening up federal R&D to thousands of small firms is valuable because it is impossible to predict from where the next breakthrough idea will come. SBIR-funded breakthroughs have often come from firms with less than ten employees.

PROGRAM RESULTS

The response to the SBIR opportunity has been enormous. More than 250,000 proposals have been submitted by firms with 500 or fewer employees since the interagency program began in 1983. More than 40,000 awards have been made to nearly 9,000 companies for a total of $6 billion to date. To demonstrate the quality of the response, many more proposals are recommended for awards than funds are available and, surprisingly, most awards have been in emerging and critical technology areas.

SBIR results to date, with respect to both the quality of the research and the commercialization, have been promising. The General Accounting Office has conducted two assessments of SBIR, the first in 1989 and the second in 1992.[1] Results in both evaluations were quite positive. GAO stated to the Congress that SBIR is a government program that works and that the program in general is doing what the Congress intended. In 1992 Congress doubled the size of the program with gradual increases to 2.5 percent of the external R&D budget of agencies with external budgets over $100 million annually and extended SBIR through the year 2000. SBIR's FY 1997 budget exceeded $1 billion.

Included in the Annex are 50 examples of the NSF SBIR research to date that have achieved commercialization.[2] They include both direct and indirect results

---

[1] See U.S. General Accounting Office, 1989, *Federal research: Assessment of Small Business Innovation Research program.* Washington, D.C.: U.S. General Accounting Office and U.S. General Accounting Office, 1992, *Federal research: Small Business Innovation Research shows success but can be strengthened.* Washington, D.C.: U.S. General Accounting Office.

[2] This research was carried out by Tibbetts in 1996.

and benefits in terms of breakthroughs, new technology and products and related sales, private investment, job growth, patents and research collaborations. Summary information for all 50 companies is shown on the attached Performance Indicators spread sheet. The 50 NSF examples were obtained through telephone and personal interviews, usually with the company president at the time of the original NSF award and through the early growth period. Often he or she was also the principal investigator on the project at the time in a very small firm. The results responded to the following questions and statement:

- Did any NSF SBIR research award make a significant difference to the performance and growth of your company?
- Did the NSF SBIR result in significant commercial sales that, directly or indirectly, you would attribute to SBIR?
- Please discuss these and other results that, in your opinion, probably would not have occurred, or would not have taken place in the same time frame, had there no SBIR program or NSF SBIR award, such as breakthroughs, sales, exports, private investment, job growth, collaborations and other benefits.
- What was the most important value of the SBIR program to your company?

Summary data on the 50 selected firms show:

- **Startup:** 16 of the 50 firms stated that the NSF projects were key to their starting the company.
- **Critical Factor:** 45 of the 50 firms said the SBIR projects were critical to their growth and/or survival.
- **Sales:** Cumulative sales directly or indirectly attributable to NSF SBIR projects total $9.1 billion. $2.2 billion was judged by the president (or occasionally another knowledgeable official) to have been directly related to SBIR research or funding from NSF. $6.9 billion was indirectly related, that is, the product did not come directly from the NSF research but would not have occurred if the NSF research or funding had not taken place because the companies attributed their start, survival, private investment, technology, spinoff or joint venture and growth to an NSF SBIR project.
- Some of the projects also involved SBIR funding from more than one agency.
- **Exports:** 34 percent of sales were exported.
- **Investment:** Private follow-on investment was $963 million of which $527 was considered directly related to the NSF projects and $436 million indirectly related. In response to the question about what aspect of SBIR was most valuable to the company, almost all responded that SBIR funded an idea for which the companies had been otherwise unable to obtain financing.
- **Jobs:** The 50 companies had a combined total of 527 employees at the time they submitted their first successful SBIR proposal to NSF. At the time of

the interview in 1996 they had an estimated 11,500 total jobs of which 9,079 were with the SBIR firm or its successor and 436 with related joint ventures or spinoffs. Net new jobs totaled 10,267. Jobs were difficult to attribute to projects and often represented company growth attributed principally to NSF.

- **Patents:** The 50 companies received an estimated 377 U.S. and 732 foreign patents that related directly or indirectly to SBIR research or funding, a total of 1,109 patents.
- **Research Collaborations:** The companies had 959 research collaborations. Of these 404 were with industrial firms, 394 with universities, 111 with national laboratories and 50 others, mostly with not-for-profit research organizations. The presidents were unable to attribute many of these specifically to individual projects. They often did say that the credibility of NSF and SBIR and the competitiveness of the SBIR program facilitated most of them.
- **Cumulative Sales:** The cumulative sales attributable directly and indirectly to NSF SBIR research since their NSF award for each of these companies ranges from $2 million to more than $2 billion. There are many other examples that were not included in which successful commercialization also has been achieved from NSF funded projects and hundreds in the case of other SBIR agencies.

NSF represents only five percent of the total SBIR program. Commercialization is in process in hundreds of additional completed projects that have had insufficient time to show significant commercial results but are expected to be marketed in the near future. Another 7,000 projects also are still in Phases I and II and many of these, too, will achieve commercial success. Many projects have not been successful for a variety of reasons, mostly because converting research-based ideas into competitive technology and commercial success is very difficult to do.

Most of the 50 companies interviewed said that they expected SBIR related new technology, sales, private investment and job growth to continue to increase each year for the foreseeable future, often at a faster rate because they had obtained private financing and market acceptance. This growth should continue in a majority of companies for a number of years.

BREAKTHROUGHS AND SIGNIFICANT INNOVATION

Many breakthroughs and significant technological innovations were made by the 50 companies. They led to commercial products and services in essentially all major areas of NSF research. These include:

- Interactive embedded software and control systems for electro-mechanical products
- All-optical switching allowing use of a high bandwidth
- Color filter and coatings for flat panel displays

- Viable non-contacting, electronic robotic arc welding sensing system
- Ultra high pressure abrasive waterjet cutting tools
- Innovative unidirectional surface acoustic wave (SAW Device) transducers
- Growing commercial tank-farmed striped bass (rockfish) on major scale
- Unmanned aircraft for atmospheric research, communications and remote sensing use
- Carbide impact bonding of aircraft turbine blades for heat and abrasion resistance
- Selective laser sintering for rapid prototyping for manufacturing competitiveness
- Ion-beam surface treatment to improve orthopedic prostheses (knees, hips, fingers)
- High resolution tomographic micro-measurement for on-line gauging of hot sheet steel
- Film sensor technology for very low temperature measurement applications
- Photochemical bonding for surface modification for biomedical and industrial use
- Supercomputer and parallel computing software that allowed their use by PC's
- Airborne optical sensors for early recognition of air turbulence and windsheer
- First long-life catalysts that allowed continuous production of monoclonal antibodies
- Critical ingredient from algae needed by nutritionally deficient babies
- World-leading handwriting recognition for pen/computer interface
- The only surgical robot cleared by FDA for operating room use (laparoscopy)
- NDE detection of glass fiber content and its distribution in reinforced plastics
- Detection of lead in paint and development of needed portable instrument
- First instrument that could reliably measure green-house gas fluxes in atmosphere
- Rapid way to identify, clone and transfer genes for (safe) bio-insecticide
- High resolution realistic full-color pictures at video-rate speeds for miniature video use
- Cross-linked enzyme catalysts (CLEC's), new technology for chemical processing
- Microwave plasma diamond deposition at high growth rates
- Revolutionary approach for teaching geometry on computers
- Membrane filtering for advanced separations of gases and liquids

- Simulation technology and operations research in production scheduling software
- NO2 detector to measure sulfur in petrochemicals, medical blood and breath.

CONCLUSION

Because SBIR companies have had the opportunity to show government agency scientists, engineers and peer reviewers what they could do, they have been able to contribute to government R&D needs, while enhancing the commercial potential of their technologies. Creative ideas and impressive research results have often led to larger Phase II awards, as well as useful collaborations and private investment in Phase III. SBIR grants have also given the company credibility in the marketplace. This, in turn, has helped them to obtain other business, investment, a larger staff and better facilities, all of which have been critical to their survival and growth. SBIR has served as a catalyst for a process which, in many cases, led to technological innovation, growth, commercialization of government research results. An important result has been a significant socio-economic return on investment from government-funded R&D to the SBIR companies, their communities, and the nation.

NATIONAL SCIENCE FOUNDATION SELECTED EXAMPLES OF NSF SBIR COMMERCIALIZATION RESULTS 1977-1995

1. Relational Technology Inc., later Ingres Corporation, Berkeley and Alameda, CA

This SBIR research project resulted in a major technical breakthrough and commercial success. It was the first distributed relational data base software that became the extremely successful Ingres software. Relational was a spinoff from UC Berkeley in late 1980 and received an SBIR award from NSF in 1981. The SBIR results led to \$1 million of venture capital in 1982, another \$2.5 million in 1984 and \$8 million in 1986 after Ingres proved a success in the marketplace. Continued fast growth led to a \$30 million Initial Public Offering (IPO). The president of Relational during this period said "SBIR was critical seed capital funding of long term strategic importance." A major contribution to Ingres' success was that "the NSF SBIR program was willing to provide early high risk funding for a very innovative, high-risk idea at the concept stage. Ingres was the first company to reach the market with distributed relational database software and it was clear that it had major growth potential in many different fields."

Customers included more than 100 of the Fortune 500 companies and included Boeing, GM, 3M, British Airways, NASA, DOD and all national laboratories. Ingres also made a significant contribution to the success of Boeing's AWACS aircraft and to airborne surveillance, particularly in the Gulf War. Ingres was acquired by ASK Computer Systems, Inc. in 1990. In 1994, ASK itself was acquired by Computer Associates International, Inc. Estimated cumulative sales that the former president doubts would have happened, or happened in the same time frame, without the early SBIR award which gave Ingres a two year lead time, now total about $2 billion. Approximately fifty percent of these sales were exports. There also have been three success story spinoffs from Ingres which in themselves have a market value in excess of $2 billion. They are Illustra, which was recently acquired by Informix, Forte Software and Documentum. The former president, now a venture capitalist, attributes this success, too, indirectly to the initial NSF SBIR award that resulted in a technical breakthrough that allowed Ingres to get the critical lead time in the field and attracted the key people who did it. Combined employment directly and indirectly related to Ingres is estimated at more than 2500.

2. Symantec Corporation, Cupertino, CA

Symantec, now a major US software company, was founded in 1982 with an NSF SBIR grant for research on the first natural language understanding (English) for microcomputers. This important breakthrough became the very successful Q&A Software. Q&A's significant sales and earnings allowed Symantec to rapidly expand and diversify through acquisitions. Symantec's founder and first president said that "the NSF project had the intellectual and commercial pizzazz that served as a magic catalyst for thinking big about the company's potential." Symantec quickly went after top quality technical and marketing staff and venture capital, and ultimately 19 acquisitions. "SBIR provided the extremely high-risk, startup financing for a very complex idea that had many exciting, potential applications and great economic leverage, if successful." It provided the early financing and profits from Q&A for Symantec to pursue rapid growth, the recruiting of 12 outstanding academic scientists and quality marketing people. Today Symantec is a major international broad-based software firm with $2 billion in cumulative sales and about 2000 employees. The success took place, interestingly, in spite of the original SBIR award being made in 1979 to another small company which was unsuccessful in its own high risk effort. However, Machine Intelligence Corporation spun off the NSF natural language understanding project with six employees as Symantec in 1982 because it believed that the idea, its potential and the early results were so promising that it could attract its own venture capital. It promptly did so and obtained $3.5 million of venture capital in 1982 followed by an IPO of $10 million in 1987. Symantec, an SBIR startup, is a world-class commercial success.

3. Flow Research, Inc., later Quest Integrated, Inc., Kent, WA

Quest Integrated, Inc. believes that it is the world leader in high pressure waterjet cutting tools and attributes this leadership to NSF and other agency funded SBIR research. NSF's Research Applied to National Needs (RANN) Program funded early research on ultra high-pressure water-jet cutting tools with Flow Research prior to the SBIR program's beginning in 1977 at NSF. In 1981 Flow's first NSF SBIR project resulted in a major improvement by adding carbide bits and other abrasives to the waterjet cutting stream which proved far more effective as a cutting tool which would then cut steel, rock, ceramics and glass. Flow quickly became the leader in the high-pressure waterjet cutting-tool field and now has an estimated 70 percent of the market. Sales from the new technology, mostly from precision metal, ceramic, glass, plastics and composite-cutting machine tools, total $250 million, made primarily by a Flow/Quest affiliated licensee. About 20 percent of sales have been exported. Major customers have included GE, Corning, Kodak, Lockheed, Rockwell and Allied Signal among others. Flow and Quest attracted $35 million in private capital and 838 new jobs have been created that directly relate to the waterjet cutting-tools technology. The companies have had research collaborations with 18 universities, five national laboratories and seven major US industrial companies. The universities included Washington, Michigan Tech, Maryland, Missouri-Rolla, Colorado, MIT and 12 others. The reader may remember the news story in 1986 when little Jessica McClure fell into an abandoned well shaft in Midland, TX. She was rescued in three hours time, after other attempts had failed, with SBIR/Flow developed waterjet rock-cutting tools. Flow Research was founded to develop new high technology products from high-risk research and spin them off as separate new companies. Five successful companies owe their origin to this approach and, in part, to SBIR. The Quest president stated that the company would not exist without SBIR due to its lack of capital and the risk involved in financing high tech research.

4. RF Monolithics, Inc., Dallas, TX

RF Monolithics, Inc. is a leading firm in the development and manufacture of advanced electronic components for the fast growing communications field. It attributes its success to its first SBIR research grant from NSF on "Theoretical Modeling of an Innovative Unidirectional Surface Acoustic Wave Transducer." The company had five employees when they spun off from Texas Instruments in 1979 and immediately prepared an NSF SBIR proposal. The single 1980 project allowed RFM to pursue "six bright ideas," all related to their next generation SAW concept. All of the ideas were technically and commercially successful. They resulted in 12 product-lines and a few hundred advanced electronic products. The company's president attributes 90 percent of their $141 million sales to date to results obtained from the theories tested under the original SBIR project, of which 50 percent of sales are exported. These include new generations of SAW devices, resonators, oscillators, receivers, filters, low-power modules, notch elements and, most recently, computer clocks. RFM focuses on the telecommunications, low-power wireless and high frequency timing markets. Major customers have included Digital Equipment, GM, Teledyne, United Technologies, Litton, Rockwell, Lucas, Siemens and Samsung. Sales have grown an average of 28 percent each year for the last six years. RFM attributes its survival and growth to the early SBIR awards that were willing to finance new concepts for emerging technology. The results helped them raise $16 million in three rounds of financing from 19 venture capital firms followed by $10 million from a public offering in 1994. Much of the money was used to build two new highly automated manufacturing plants that use flexible manufacturing processes. These have significantly contributed to the company's growth and competitiveness. RFM has received 78 patents and has collaborations with the University of Maine, UT Dallas, Central Florida and Chiba in Japan.

5. Aquatic Systems, Inc., now a division of Kent SeaFarms, Inc., San Diego, CA

Aquatic Systems Division of Kent SeaFarms is the producer of more than half of all striped bass (rockfish) sold commercially in the US as a result of the NSF SBIR research and the program's incentive to commercialize the results. In 1982 SBIR provided the first R&D funding for two marine scientists who had spun off as a consulting firm in 1972 from San Diego State University. ASD's president said that the SBIR goal to commercialize research results stimulated them to convert their small consulting firm into a commercial fish-farming company and to focus Aquatic Systems on a business growth strategy. The SBIR research involved the effect of water temperature and quality, and various feeding combinations, on fish growth rate. The initial NSF research project proved that they could produce 50 times the number of pounds of fish per acre as the average production of 3500 competitors using conventional fish farming methods. Their success attracted another $6 million from Kent SeaFarms to significantly expand their tank farm facilities. Today, Aquatic Systems has 98 large storage tanks containing a few million fish located in the desert east of San Diego. It is now a large fish farming company and is exploring other US and foreign sites for the production of striped bass and other species of fish. Cumulative commercial fish sales are $33 million and employment is up from 7 to 70. SBIR research results and the credibility of NSF helped the company to obtain $2.6 million in limited partnership funds. Research collaborations include San Diego State, North Carolina State and Clemson universities, supplier firms, NOAA's Southwest Fisheries, and our National Marine and Canadian Marine Laboratories. The company is also conducting research on the use of their waste water for agricultural and environmental advances as an outgrowth of the SBIR research. ASD has received an ATP award relating to water quality treatment as well.

6. Collaborative Research, Inc., now Genome Therapeutics, Inc., Bedford, MA

Collaborative was a pioneer US firm in the genetics field thanks to NSF's first SBIR genetics grant in 1977 which the founder, former president and chairman, said "changed the history of the company." The timing for a genetics company was perfect with Genentec having been founded the same year. Collaborative's genetics award from NSF, the possibilities for the emerging genetics field and the commercialization objectives of SBIR research led to a large investment by Dow Chemical. This attracted other venture capital; a total of $36 million of equity funding was obtained by a company that had 33 employees at the time of the Phase I award. Collaborative was a pioneer in the genetics field. It cloned Interferon, bovine enzymes and renin and made a major contribution to the genetics field by being the first to complete and publish a genetic linkage map of the entire human genome in 1987 using $5 million of its own funds for the research. This risk paid off as the mapping became a powerful tool for drug development. It also helped Collaborative to obtain a $37 million grant from NIH, the largest private sector award for a human genome project. Collaborative Research was the first to identify the cystic fibrosis gene at chromosome 7 and has conducted more than 50 genetic probes. The drug development initiative attracted two major drug firms, Astra and Schering Plough, with long-term contracts totaling $65 million. Pfizer is now the largest producer of renin in the world through its license of a Dow-assigned Collaborative patent. Non-genome Collaborative business was sold off gradually to other companies and in 1994 the company became Genome Therapeutics. Cumulative sales, directly and indirectly related to the NSF funded, first genome project, and later NIH, exceed $50 million. Job growth in Collaborative/Genome Therapeutics and in firms spun off from Collaborative total 227. The company has collaborated with more than 50 medical scientists in many research institutions all over the world. These include much collaboration with Harvard and Massachusetts General Hospital and, in the cystic fibrosis project, the Universities of Michigan and Toronto.

7. Advanced Technology Materials, Inc., Danbury, CT

The potential of the small high tech firm to quickly convert research into economic results was demonstrated by Advanced Technology Materials which was founded in 1986. It has grown rapidly to 170 employees as a result SBIR funded R&D and their focus on commercial markets. It has had two major innovation successes, one in micro electronics vapor deposition and the other in environmental control equipment which it attributes to NSF, DOD and EPA SBIR research. Advanced Technology Materials submitted a proposal to NSF soon after its founding for "Molecular Chemistry of the Superconductor Substrate Interface," a fairly basic and key project in thin films. The promising results led to $250,000 from Millipore and then $1 million in venture capital after some early difficult times where NSF and DOD SBIR funding was critical to ATM's survival. The NSF work was in a hot area that resulted in a 50 percent growth rate for three years. ATM used the technology to explore a new approach to producing DRAMS and won a $10 million contract funded half by DOD/ARPA and the other half by IBM, Texas Instruments and Micron Semiconductor. The three companies will manufacture the DRAMS, if the ATM technical approach is successful. Other thin film sales total another $10 million and an EPA SBIR has resulted in $30 million sales of environmental control equipment for semiconductor manufacturing. Other major customers include Motorola, HP, TI, Intel, Digital, Samsung and Hyundai. About 40 percent of sales are exported. The company has raised $47 million, with $9 million of venture capital, $30 million in public offerings and $8 million from industrial investors. ATM has 110 issued patents and another 190 have been applied for that relate to SBIR funded research from a number of agencies. The company has used more than 50 university consultants, with emphasis on MIT and Yale, but which also include Stanford, Caltech, UCLA, UC Berkeley, Cornell and RPI. It has also collaborated with more than 100 large companies on contracts including IBM, Texas Instruments, Hewlett-Packard, Varian, Siemens and Micron. ATM's growth rate in two important technology areas has resulted in a company (stock) market value of $100 million.

8. Aurora Flight Sciences Corporation, Manassas, VA

Another example of early SBIR conversion of research into products is Aurora Flight Services which was founded in early 1989 and received two NSF awards in January 1990. The first research was to design an innovative unmanned aircraft "Theseus: A New Platform for High Altitude Atmospheric Science." The other related to developing a "Lightweight Dropwindsonde System for Unmanned Aircraft." Both projects were technical successes and attracted $4.5 million additional venture capital following the initial startup capital of $200,000. Product sales to date already total $30 million. The company is expecting additional sales of $50 to $100 million in a year or two from outstanding aircraft proposals for use in such areas as communications, defense, atmospheric sciences and remote sensing. The latter proposed use is from foreign countries. The success of Theseus led to DOD funding a larger plane called Perseus and its purchase by NASA, DOE and others for a total of $21 million. Aurora employment has increased from 3 at the time of the 1988 NSF proposals to 94. Extensive collaboration has been with 6 universities: Harvard, MIT, VPI, the University of Virginia, and West Virginia and Old Dominion Universities. Federal laboratories have also been involved in collaborative research with Aurora, including Sandia,

Los Alamos, NCAR, and 6 NASA centers. Aurora says that NSF credibility has been important to their success everywhere they have gone, particularly in marketing their technology in the US and internationally. The NSF SBIR also played a key role in providing critical early funding.

9. Browning Engineering Co., Hanover, NH

Browning is an example of SBIR's support of an individual inventor who licensed his SBIR developed radical plasma-jet spray bonding technology that is being used by all major jet aircraft engine manufacturers to strengthen the resistance of compressor blades to extremely high temperatures, abrasion and corrosion and to increase stiffness. These sales and others have resulted in Browning receiving more than $3 million in royalties. The research was to pursue innovative ideas in "Extreme Impact Velocity Metal and Ceramic Deposition" for bonding high density refractory layers of materials, such as tungsten carbide, into a stainless steel surface and takes place at more than twice the speed of sound. Other materials such as ceramics, stainless steel and zinc can be applied as corrosion resistant coatings. Nelson Browning spun off from the Dartmouth College faculty prior to SBIR but he was unable to pursue his innovative idea until receiving the NSF SBIR award. The technology is also used on steel rollers for producing other steel rollers, for steel and paper production and by the printing industry. Licensed total sales to date are about $50 million of which 30 percent are exported. Users include GE, the Pratt & Whitney Division of United Technologies, Rolls Royce, SNECMA (French aircraft engine producer), the Allison Division of GM and numerous smaller users. Browning has 35 SBIR related patents and about 125 overall. University collaboration has been principally with Dartmouth and MIT. Dr. Browning also is a recognized leader in the hyper velocity oxygen fuel field (HVOF). He recently received an award at ASM International's annual meeting for his significant contributions to thermal spray technology. He also has recently been made one of only four members of the Jetspray Hall of Fame. Browning Engineering previously carried out another exciting NSF (non-SBIR) project which involved flamejet ice drilling where he drilled a needed 3 foot hole through 1428 feet of the Ross Ice Shelf in 7 hours to allow Antarctica research projects under the ice.

10. Nova Automation Corporation, later DTM Corporation, Austin, TX

Nova Automation spun off from the University of Texas at Austin to pursue possibilities for laser sintering technology in the rapid prototyping field. It immediately submitted an SBIR proposal for " Selective Laser Sintering." Today, after a name change, DTM is a world leader in rapid prototyping. The process is improving product design and activity, shortening the manufacturing process, lowering costs and accelerating the product's speed to market. The commercial success was the result of a spinout of university research and a single Phase I award. However, the company states that the SBIR award was critical to the success in a number of ways. Most directly was the early financial survival of the company as it was the source of the first outside financing when no other financing was available. NSF credibility helped NOVA gain cooperation and a licensing agreement from the University of Texas. The NSF SBIR grant facilitated a UT scientist's receiving a $400,000 grant from the state to continue his research and collaboration with the company. The SBIR research results leveraged Nova's obtaining $3 million in venture capital and later, substantial additional investment from BF Goodrich. The NSF and SBIR credibility assisted Nova in being selected as the first beneficiary of the new UT Center for Technology Development and Transfer, established to increase the conversion of university research into commercial technology. Upon the completion of Phase I the company obtained its follow-on funding commitment to support Phase III (if Phase II was successful) from BF Goodrich. As BFG became more familiar with Nova, the technology and its commercial potential, it invested more money in Nova and shortly thereafter, prior to NSF's Phase II award, bought control. This made Nova, soon to become DTM, ineligible to receive the Phase II award. Today DTM is a world leader in the rapid prototyping field with its computerized sintering machines that produce three dimensional models, prototypes and parts. Total sales are estimated $40 million and private investment around $43 million. Employment has increased from 4 to about 100. There are about 27 related patents. Close collaboration with the University of Texas continues.

11. Spire Corporation, Bedford, MA

In 1979 Spire received an NSF SBIR award for"Improved Wear Resistant Titanium Alloys for Orthopaedic Prosthesis". Today Spire is the principal provider of the advanced ion-beam surface treatment to improve the wear and infection resistance, bio-and hemo-compatibility and slipperiness of artificial knees, hips and other implanted prosthetics as a result of the project. Spire also produces other products in the materials, electronics and process equipment fields that have benefited from NSF and other agencies' SBIR funded research. Artificial joint sales total over $30 million and a spinoff another $6 million. Principal customers include Bristol-Myers, Johnson & Johnson, and Stryker Corporation, and Kennemetal and Caterpiller for other NSF projects applications. Employment has increased from 90 at the time of the first NSF SBIR award to 150. There are 27 SBIR related patents, 8 of them from NSF projects. Spire has had more than 130 research collaborations, many of them related to NSF SBIR projects. They include MIT, Harvard, Caltech, Brown, Boston University, Los Alamos, Livermore, WPAFB, Argonne, Sandia, Motorola, Norden, Raytheon, Hughes, Texas Instruments, Rockwell, Lockheed and Caterpiller. SBIR provided the R&D funding for Spire to pursue essentially all of the research behind their current products because of the degree of risk involved. The president said, "Without SBIR the company probably would have survived but at a much smaller size."

12. Scientific Measurement Systems, Inc., Austin, TX

The company is a leader in the design and manufacture of industrial tomographic systems as a result of NSF research projects on high-resolution tomographic measurement for on-line instant thickness gauging of hot sheet steel. This breakthrough accomplishment involved collaboration with Bethlehem Steel Corporation that resulted in SMS' innovative gauging equipment being installed at Bethlehem's largest sheet steel plant at Sparrows Point, MD. Bethlehem states that the plant produces the world's highest quality sheet steel. This is of major importance to the US automobile industry as thinner steel has been required to reduce the weight of today's automobiles requires the close thickness tolerances that the SBIR research made possible. SMS tomographic gauges measure the thickness of every inch of the entire sheet surface at production line speeds to within 4 microns of accuracy and allows for instant on-line adjustment of rollers to correct variations. In addition to gauging systems, SMS produces small, medium and large object resolution tomographic scanners, most of it for NDE applications. Sales directly related to NSF SBIR research total $30 million. SMS is a member of a team that won a large automobile related ATP project as a result of its SBIR funded research base and experience. Major SMS customers include US Steel, General Motors, United Technologies, Allied Signal, Rocketdyne, Garrett, GE, NASA, Oak Ridge, and British Petroleum. Twenty percent of sales are exported to Canada, Japan, Germany, Switzerland, China and Italy. SMS spun off from the Nuclear Physics Lab at the University of Texas. Collaboration has been with the Universities of Texas and North Texas, Bethlehem, GM, GE, and EG&G. The company has 4 patents. It also states that it survived in 1981-1982 due to SBIR funding and later obtained $8 million of outside financing largely based on the success of the SBIR technology. The number of jobs has increased from 13 at first NSF SBIR award to 25.

13. IDM Corporation, Austin, TX

This is an example of how SBIR research results can directly benefit more than one company. IDM did not receive an SBIR award but its existence is a direct result of the SBIR program. IDM is a spinoff from Scientific Measurement Systems using the same NSF SBIR funded technology, high-resolution tomographic measurement, for other applications. IDM sales total $16 million. It has had investments of $4 million. The founder, president and principal investigator of SMS left by mutual agreement between the companies to found IDM. IDM pursues the technology into other markets such as the petrochemical, power utilities and the pulp and paper industries. New IDM products were developed for quality assurance and quality control using similar gamma ray gauging systems, process control and monitoring equipment, as well as mobile equipment for field-use. Sales have all been based upon modifications of the Scientific Measurement Systems technology developed from the same SBIR research. Major customers are Mobil Oil, Texas Utilities Electric, Florida Power & Light and Boston Edison. Employment from startup in 1988 is now 28. IDI has 91 related patents. Research collaborations have been with the University of North Texas and research consortium of IDM, the Electric Power Research Institute and Florida Power & Light, to pursue power-company related non-destructive evaluation (NDE) applications.

14. Bend Research, Inc., Bend, OR

Bend is a leading small company in developing synthetic membranes for filtering applications and advanced separation systems for gases and liquids. NSF SBIR results led to separating ions from mixtures to 99 percent purity, membrane reactors to purify sweeteners, and separating 95 percent of oxygen from air. Cumulative sales total estimated at $90 million with $23 million directly attributable to NSF projects. Major customers have included Pfizer, WR Grace, Coca Cola, Lilly, Teledyne, Merck, J&J, Exxon, Gulf, Chevron, Unical, Mobil, Nippon Santa, Nippon Kokim, and Elf Atochem. Bend has had 4 spinoffs, three of which have been successful— Consep, was a joint venture with Bethlehem Steel; Grace Membrane Systems a joint venture with WR Grace; and Bend Research Manufacturing, Inc is a wholly owned subsidiary. Bend has grown from 10 to 75 employees and another 160 jobs have been created in the spinoffs. Combined private investment totals $48 million which Bend attributes to its substantial research base in emerging technology areas. This has been created through more than 150 SBIR awards over 17 years from many agencies. The company has 22 patents and 12 research collaborations with the universities of Oregon, Oregon State, Portland State, UC Berkeley, UC San Diego, Minnesota and Colorado and also with Pfizer, WR Grace, Teledyne, Western Gas and Sandia. Bend attributes its success to NSF's willingness to fund high-risk, innovative ideas in the emerging separation and controlled release field, which provided half the company's modest income in early years and later the funding of many other high risk, R&D ideas by many SBIR agencies.

15. Integrated Systems, Inc., Santa Clara, CA

Integrated Systems is possibly the world leader in interactive embedded software and control systems design and services and has been a pioneer in the field. ISI was founded in 1980 and received an NSF SBIR award for research on "Computer Aided Design Methods for Systems Analysis" in 1981. The interactive embedded software greatly improves product design for manufacture and use, provides critical interactive control options and can significantly reduce the time to get a product from design to market. Integrated's products are used extensively by the automotive, aerospace, defense, office and factory automation, telecommunications, process control and now the multi-media industries. Examples include their embedded chips in automobiles and in hundreds of mechanical products from the space shuttle to small appliances. They include the use of the credit card to automate payment at the gasoline pump which proved time-saving, increased station efficiency, provided better controls and which has been popular with consumers. The embedded software concept came from the founder's doctoral thesis at Stanford. He said that "NSF's first project took the critical risk to do the required research that we otherwise could not have taken, at least at that time, that was critical to the company's success." Today the company's clients read like a who's who of major manufacturers in large and emerging industries. Cumulative sales are all attributed, directly and indirectly, to the initial breakthrough concept funded by NSF and later to other agencies for more specific applications. Sales now total $250 million, with one-third exported. Major clients include Motorola, Ford, AT&T, GE, Kodak, GTE, Hewlett Packard, Boeing, Lockheed, McDonnell Douglas, Commonwealth Edison, Northern Telcom, Fujitsu and Alcatel. Substantial SBIR funding has also come from DOD for military applications. Employment has increased from 11 at the time of NSF's first SBIR award to 450 at present. The company has collaborated in its research with Stanford, Caltech, MIT, Novell, Boeing and Sandia. It has 8 patents. The founder also stated that "Integated's very existence and growth are directly attributable to the SBIR program."

16. LakeShore Cryotronics, Inc., Westerville, OH

The NSF SBIR in 1981 award to LakeShore Cryotronics was the beginning of the company's development of thin film sensor technology for low temperature metrology applications. Today the company believes that it is the world leader in this field and in the use of temperature measurement in high magnetic fields under ionizing radiation. It is also strong in low-temperature physics sensors and instruments for research, testing and industrial use. All of this capability is directly related to the development of ideas for the use of thin films as temperature sensors under the SBIR research grant. The products and related services represent 40 percent of the company's business. LakeShore attributes its success in this field to NSF's willingness to fund the high risk research required that allowed it to focus on a new potential growth area. Total sales from the three product-lines, low temperature sensors, low temperature measurement control instruments and magnetic property characterization systems, total $30 million, of which they attribute about $20 million to NSF funded ideas. About 45 percent of sales are exported. Major clients include CNRS, Max Planck Institutes, DOE National Laboratories, IBM, Hughes, GE, Motorola, McDonnell Douglas, Seagate and Martin-Marietta. Foreign sales are primarily to Europe and the Pacific Rim. There are two important NSF related patents. Research collaborations have been with Motorola, Micron, the following universities: Ohio State, Michigan, MIT, Brown, Johns Hopkins and Tokyo; with Argonne and Los Alamos National Laboratories; and with an Ohio State consortia of 200 companies. Employment has increased from 45 at the time of the first NSF award to 110 at present.

17. BioMetric Systems, Inc., later BSI Corporation, Eden Prairie, MN

Biometric Systems, Inc. spun off with three people from another small company in 1979 with an NSF SBIR project and a small contract with the FDA as its only business. The project was for " Synthesis of New Photoreagents for Photochemical Immobilization" Today BSI believes it is the world leader in using photochemical bonding for surface modification of biomedical and high value products and industrial processes that began with this first NSF project. As the company grew it obtained $13 million in private investment it attributes to SBIR, about half of it to NSF. Sales by licensees are estimated at about $70 million directly related to SBIR, with $35 million related to NSF. About 20 percent of sales are exported, mostly to Germany and France. (There were also significant indirect benefits of SBIR to the company such as another $130 million of licensed sales made by the company that would not have been made had the company not survived.) Company employment, which relates principally to SBIR funded R&D, has increased from 3 to 73. BSI has produced 59 SBIR related patents. Research collaborations include the following universities: Minnesota, Harvard, Michigan, Wright State, Missouri, Wake Forest, Utah and East Carolina. Industrial collaborations have been with American Cyanamid, Ciba-Geigy, Corning and Bosch and Lomb. Other collaboration has been with the Naval Research Laboratories.

18. Decision Science Consortium, now Decision Science Associates, Manassas, VA

Commercial success in SBIR has also occurred in the social and economic sciences. Decision Sciences Consortium received NSF awards in 1983, 1984 and 1985 for Role Specific Judgment Biases in Decision Making; Analysis Methodology for Measuring Populations of Institutions; and Integrated Framework for Knowledge Representation and Acquisition. SBIR changed the focus of the company's research from theoretical behavior and economic in nature to an emphasis on more practical objectives to help leaders who make decisions on government and commercial problems be more responsive to public and company needs. Decision Sciences was successful in developing this change in focus and provided personal, strategic and policy neutral services at high level meetings of decision-makers on critical problems. The goal was to help keep the focus on practical company objectives and needs, and equally on how to achieve them.

These included meetings at the White House, Defense Department, Nuclear Regulatory Commission and in the international area, such as COCOM Commission meetings on export controls on Eastern Europe. The commercialization objective of SBIR resulted in the company finding a Phase III partner, ICF, later IC Kaiser, which purchased Decision Sciences for $3 million. Commercial clients included IBM, Ford Motor and Cutler Hammer. Collaborations included the following universities: Pennsylvania, Dartmouth, Harvard, London, Stanford, Michigan State, George Washington. Others were with Argonne, Los Alamos, Lawrence Livermore and Berkeley, Battelle NW, Sandia and Brookhaven. Employment increased from 15 to 35 at the time the company was sold.

19. Charles Evans & Associates, Redwood City, CA

Charles Evans & Associates had 10 employees at the time of their first NSF SBIR award in 1981 for research on "An Advanced Detection System for Quantitative Three-Dimensional Ion Microscopy." Through many SBIR awards from a number of agencies, particularly NSF, NIH and DOD the company has been a very successful designer and manufacturer of state-of-the-art mass spectrometry systems (SIMS) and related devices. The company has received three R&D 100 awards and is an internationally recognized leader in its scientific instrumentation field. The founder stated that "It wouldn't have happened without SBIR." SBIR R&D supported idea to prototype funding, which has been critical to the company's success. Cumulative sales are $45 million with about one-half attributable to NSF and 50 percent exported. The company has had over 2000 customers that include IBM, AT&T Bell Labs, Intel, NIST, dozens of universities, Sandia, Livermore, Toshiba, Nippon Steel, Matsushita, Sony and many other name research institutions. Foreign sales have been mostly in Europe and Japan. The company has 5 US patents and doesn't obtain foreign patents because foreign instrument firms do not build instruments for foreign markets only. CEA has had more than 100 collaborations with universities and has made many research grants to students. These included Illinois, Cornell, Caltech, Wisconsin, Paris, Duke, Utah, Illinois-Chicago and the UC San Francisco's Medical School. Collaborations have also been with Sandia, Pacific Northwest Laboratories, Oak Ridge National Laboratories and with many name foreign research institutions. Employment has increased from 10 to 100 plus an additional 28 who were in a sold-off division.

20. Pritsker & Associates, Inc., Indianapolis, IN

Pritsker & Associates developed production scheduling software from a 1981 NSF project entitled "Development of Operations Research Algorithms and Support Data Structures for Microcomputers." Two of the three product-lines, FACTROL and Simulation, have a number of modules for selective applications and are used in 250 manufacturing sites worldwide. They represent half the company's revenue. The third product line is a Hospital Operating and Recovery Room Scheduler. Pritsker is a former professor of engineering at Purdue University, who left Purdue to devote full-time to his first SBIR project. Pritsker was one of the first to apply simulation technology to develop operational manufacturing schedules through the use of operations research algorithms and later expert systems. Direct, indirect and licensed sales total $70 million with 30 percent exported. Clients included Caterpiller, IBM, Reynolds Metals, RL Donnelly, Eli Lilly, Boeing, Bethlehem Steel, GE, Motorola and many others. Pritsker is proud of his company's development of the Organ Network for Organ Sharing software. FACTROL is used the world over by clients in Japan, Germany, Korea, UK, Canada, Mexico, Brazil, South Africa, Taiwan, Turkey, Singapore, etc. The SBIR support led to the company's obtaining $1 million in venture capital in 1992. Collaborations include Purdue, Northwestern, Georgia Tech, IBM and Caterpiller. Employment increased from 6 to 140 in 1991 before industrial downsizing began which has resulted in a steady reduction in orders and in staff to 50.

21. Helisys, Inc. (formerly Hydronautics, Inc.), Torrance, CA

The founder and president is a recent Russian immigrant to the US and set up his company to commercialize some his innovative ideas in rapid prototyping processes. He founded Helisys (as Hydronautics) in 1988 and immediately submitted an SBIR proposal, which was funded by NSF in 1989. Helisys designed and built an automated Laminated Object Machine (LOM) and obtained one of the first patents for rapid prototyping/digital manufacturing by converting digital information into a 3D physical prototypes using a laser to fuse powder one layer at a time. The use of sheet materials for making dies and molds has been a second significant innovation that has further reduced cost and simplified the process. The company has been very successful commercially. Sales total $33 million and reached a new high of $13 million for the year ended July 31, 1996, up 25 percent from the previous year. Helisys has two principal products, a small machine that sells for about $100,000 and a larger one for about $200,000. LOM technology can often convert ideas into prototype 3D hard objects in a few hours, can significantly reduce development cycles and the time and cost to bring new products to market. The company now has 102 employees. The president said, "When we made our first beta system in 1991 we were just about broke. No venture capitalist was willing to take the risk. NSF was willing to fund a radical idea and we made our breakthroughs with SBIR funding." About 45 percent of sales are exported to Europe and Japan. Much of their business is through service bureaus to small companies. Major customers also include GM, Ford, Mercury Marine, Dana Corporation, Mercedes, Fiat, Baxter, Rubbermaid, Boeing and Hughes. Helisys has 5 patents and has recently raised $7 million with an initial public offering. Collaborations are with the University of Dayton and UCLA.

22. Scientific Computing Associates, Inc., New Haven, CT

Scientific Computing Associates has been a pioneer in high performance software development for supercomputers and parallel programming. It collaborates closely with Yale and states that it produced a number of breakthroughs in commercial parallel computing software from NSF SBIR awards. These include important innovations such as: "Linda" the first robust, portable codes for the first parallel computers; "Network Linda" which was the first to use the idle cycles on workstations; "Paradise" that allowed the use of the supercomputer by smaller computers down to PC's; and software to allow the writing of parallel codes in diverse disciplines, such as quantum chemistry and sophisticated financial analysis, with real-time programming to PC's. Company sales to date are $21 million, about half in contract R&D and half in software, both of which were helped a great deal by NSF's credibility as well as funding. Major customers are universities, national laboratories, National Computing Centers, and world market companies such as American Express, Wall Street banks, AT&T, Boeing, Lockheed, Exxon, Shell, Chevron, IBM, Apple, Motorola, Westinghouse, Siemens, Schlumberger, Glaxo, Hitachi, Daimler-Benz, Fujitsu and dozens of others. SCA also has joint marketing agreements with DEC, IBM, Cray, and HP/ Convex. The president said, "We would not be here today without SBIR. SBIR was willing to take the high risk of designing software for the supercomputer field, particularly for commercial applications." Employment has increased from 8 at the time of first award to 15. Collaborations include Yale, Maryland, Princeton, Harvard, Duke, many oil companies, financial firms, Motorola, General Dynamics, Argonne and Sandia.

23. EPITAXX Inc., Princeton, NJ

EPITAXX , a spinoff from the now-closed RCA Laboratories, was founded in 1984 and received an NSF SBIR award in 1986. Its president said that by 1990 EPITAXX was the foremost company in the world in indium gallium arsenide detector technology for special applications, such as fiber optic cable TV detectors and near infrared detector arrays for satellite imaging. A second NSF award produced a device that led to the next generation optics technology on a chip. These were significant technical achievements, particularly for a new firm. The company also received SBIR's from DOD and NASA and grew rapidly but without sufficient earnings to attract US investors. No American firms responded to a request for offers although two UK companies and a Japanese firm did. EPITAXX accepted a $12 million offer from Nippon Sheet Steel which would keep the same staff and provide financing. A year later, however, the Japanese firm put in its own president. EPITAXX continued to grow and, in a change in strategy, increased its emphasis on manufacturing versus R&D. Cumulative total sales are estimated at $76 million, about half of which were attributed to SBIR and half of that, or $19 million, directly to the NSF funded research. Major customers of EPITAXX products were HP, GE, Siemens and Oki. About 60 percent of sales were exported. Employment increased from 5 at the time of the early NSF award to 160 today plus 22 in the original founder's spinoff company, Sensors Unlimited, which is already an NSF SBIR firm. Collaborations were with the University of Southern California and the New Jersey Institute of Technology.

24. Crystal Systems, Inc., Salem, MA

Crystal Systems was founded in 1971 as a spinoff from the Army Materials Laboratory in Watertown, MA. The company received NSF SBIR awards in 1979, 1982 and 1983 related to growing pure crystals of ruby, titanium doped sapphire and cobalt doped magnesium. The titanium-doped sapphire started a new class of extremely high quality, electronic crystals that had a wider range of tunability that made tunable lasers far more feasible than ever before. Crystal Systems' titanium sapphire crystals are the highest quality and most efficient laser crystal available and were key to the development of the first, commercially successful, tunable lasers. Most tunable lasers throughout the world today use these crystals made by Crystal Systems that came from the NSF SBIR funded research. Although the actual crystal sales of the company to date are only $5 million, the company states that the economic payoff has been that the purity of the NSF/Crystal Systems crystal made the tunable laser technically and commercially viable. CS says that most tunable laser manufacturers use this crystal today. Major customers of CS include Coherent, Spectra Physics, and Laser Photonics. Cumulative estimated indirect sales (of CS tunable lasers) total about $100 million. Only about 10 percent of the crystals are exported as most tunable lasers are made in the US. The company has 10 issued patents. Collaborations include MIT, Michigan, New Mexico, UC San Diego, Coherent, Spectra Physics, Schwartz Electro-Optics, GE, GTE, Lockheed and Lincoln Lab, Livermore, Oak Ridge, Los Alamos and Sandia National Labs.

25. Weidlinger Associates, Los Altos, CA

Weidlinger is principally a mathematics modeling firm for geological investigations related to oil exploration. SBIR provided the research opportunity for the company's 3D modeling techniques to explore new ideas in these and other areas, such as ceramics and electronics. The results of 3D simulation research may have revolutionized oil drilling. The computerized 3D modeling of a geology system is possibly the most versatile and best techniques in the world because it can handle both inverse and forward modeling with wave simulation for use in locating oil, seismic earthquake activity, NDE for nuclear reactors and other tailored user needs. Based on other NSF awards to Weidlinger, it also can simulate the response of a piezo-electric transducer 100 times faster than other approaches for electronics applications. More than 150 licenses of the software have been sold for modeling oil pools world-wide and these have and will continue to increase oil reserves, often significantly, as in the recent major deep water discoveries in the Gulf of Mexico. Another NSF award has focused on application of the modeling technique to another area to increase the fast simulation of response of piezo transducers, which also may be very important. A company official said, "SBIR has funded most of the research the company has done in fundamental areas which has revolutionized the generic tools of computerized modeling for many applications. It has resulted in a major successful change in the direction of the company." Directly related income from software royalties and contracts alone total $6 million. Employment in the company, partially related to SBIR, has increased from 90 to 200. Collaborations include Stanford, UNC, Arizona State, Strathclyde, UCLA, Washington, Livermore, Argonne, GX Technology (oil exploration), Schlumberger, Amoco, EPRI and SEMATECH.

26. Terra Tek, Inc., Salt Lake City, UT

Terra Tek is a leading petroleum related R&D firm world-wide, particularly in the drilling area. NSF SBIR research has made significant contributions to their success by funding early stage ideas, in both the service and product fields, that often needed to show promise between the idea and a prototype before oil industry support could be obtained. Terra Tek received an SBIR award in the first SBIR program solicitation for proposals at NSF in 1977 on the need for Simplification of Methods for Measuring (Metal) Fracture Toughness. The research was successful and a Fractometer instrument was being sold to the oil drilling industry world-wide prior to the completion of Phase II. Sales to the industry to date are now $5 million. In 1988 the project was on X-Ray Computerized Tomography Applications to Ceramics and Composites. It resulted in a new way to x-ray and assess the quality of coal relative to shale content. Sales of x-ray systems software and hardware already exceed $15 million and will lead to a generation of new geotechnical applications. In 1990 the project was on Self-Stabilizing Underground Excavation Cross Section and it will help increase the permeability around the well-bore and service and equipment business. Sales to date are over $2 million. These and other SBIR awards from NSF and other agencies have resulted in related R&D business from the oil industry to date that totals $13 million. Total NSF related sales, directly and indirectly are approximately $37 million. Customers include most of the world's major oil producers— in North, South and Central America, Europe, the Near East, Africa, the Far East, Southeast Asia and Russia. SBIR has helped the company raise $35 million, about one-quarter attributable to NSF. NSF funding has led to 12 patents. Employment increased from 85 to 400 at one time prior to 3 spinoffs resulting in a decrease to 60. Research collaborations have been extensive and include Utah, Brown, UC Berkeley, Caltech, Cornell, Texas A&M, Stanford, MIT, Colorado Mines, Missouri-Rolla, BYU, New Mexico Tech and King Saud (Arabia). It also has been with all 9 DOE national labs and with an estimated 100 companies including most of the world principal oil companies and also GM, Caterpillar, TRW, Halliburton, Baker Hughes and Burlington Northern. Foreign collaborators have included Mitsubishi, Sumitomo, China Petroleum (Taiwan), China National Petroleum, Nigerian Oil, Venezuela Interbep and Birmigaz (Russia).

27. Advanced Research and Applications Corporation (ARACOR), Sunnyvale, CA

ARACOR was the first to build an industrial computed tomographic scanner as a result of early NSF and DOD, NASA and DOE SBIR research. They have had many firsts in the field including designing and building the world's (1) largest, (2) highest energy and (3) probably highest resolution scanners for use in inspecting nuclear missiles, rocket motors and related weapons, parts and equipment. The scanners have saved millions of dollars of weapons and equipment by inspecting them for QC purposes, including for radiation hardness before they are sent into space. Currently they are involved in inspections relative to the reduction of the number of nuclear missiles. There have been other civil sector payoffs. The SBIR research has led to the development of a similar but lower cost radioscopic inspection systems, 55 radiation hardness quality control tester systems for parts to be used on satellites or otherwise going into space, and an X-ray fluorescent spectrometer for use on the coming NASA Mars mission. It has also generated related software, spun out a subsidiary for manufacturing bonded wafers and initiated a joint venture with Metorex (Finland) to produce layered microsructures (LSM's) and a new analytical scanning electron microscope is under development with Intel. NSF related sales total $82 million with 5 percent exports, mostly to France, the UK and Singapore. Major customers have been Defense, NASA, DOE, Aerojet, TI, Hughes, TRW, GE, Motorola, Lockheed, Honeywell and Howmet. Employment has increased from 25 at the time of the first NSF SBIR award to 48 today. There are 21 related patents and collaborations have been with Cornell, Rochester, Stanford, Southern California, Lawrence Livermore, Lawrence Berkeley and Sandia Laboratories and with Aerojet, Howmet and Intel. ARACOR is now a team member on a joint DOD project GE, Hughes, AT&T and Howmet.

28. Radiation Monitoring Devices, Inc., Watertown, MA

Radiation Monitoring Devices has converted three NSF SBIR projects into three separate, innovative and commercially successful instruments in three different fields, detecting glass content in reinforced plastics, lead in paint and freeze measurement in production processes. They have made significant contributions to quality control of the distribution of glass fibers in strengthening plastics, to public health in detecting lead in paint and the need to detect and measure freezing in production processes. An NSF SBIR grant in 1982 for Rapid, Non-Destructive Measurement of the Glass Content of Reinforced Plastic Composites RMD resulted in Compuglass, the first instrument to be able to measure the content and distribution of glass and other additives in plastic and composites. The non-destructive analysis is carried out in real-time, including large samples on production lines, such as in the automotive parts and the exploding office machine and electronics field of casings and parts. A 1985 grant was for Lead Iodide Semiconductor Nuclear Sensors was one of the first to detect lead in paint and in oil well logging. In 1990 RMD proposed a new concept to develop a Gradient Freeze Method Instrument. The freeze measurement instrument is important for food and other production processes. RDM stated that SBIR funded their high-risk, research ideas on projects that otherwise could not have been undertaken, the ideas were critical to the company's growth and SBIR converted RMD into a product company rather than solely contract R&D. As to results, RMD has received three R&D 100 awards, direct NSF related sales total $12 million and those from their research base another $8 million. About 15 percent of sales have been exported. Customers include GM, DuPont, Ford, Rockwell, Eagle-Picher, Ferro, Reynolds, BASF, Thomson, Bayer, Ciba-Geigy, Glaxo and AKZO. Employment since first NSF SBIR award has increased from 11 to 43 today. SBIR related collaborations, principally relating to NSF and NIH funded research, total 34. They include NJIT, MIT, BU, Stanford, Michigan, Johns Hopkins and Harvard; also Schlumberger, Xerox, Fisher Medical and Canberra. National Laboratory collaborations have been with Lawrence Berkeley, Sandia and Los Alamos.

29. The Merrick Corporation, formerly in Nashville, TN

Merrick made a major, innovative breakthrough inventing the world's first viable, non-contacting, robotic, arc welding sensing control system during an NSF SBIR project in 1980. It is estimated that more than 75 percent of all arc welding robots used in the world since 1984 use some version of the sensing system developed under the NSF project. This ThruArc electronic sensing system made a major contribution to the robotic welding field and the project's principal investigator, Dr. George Cook, received the James F. Lincoln Arc Welding Foundation Gold Medal in 1981 for his contribution to the industry. NSF's SBIR program was the first to provide any significant funds to the company and this allowed them to pursue their idea that radically changed arc welding. It was a system that can automatically sense the joint electronically and alter the programmed path of the robotic manipulator in real-time to weld and to avoid inaccuracies. Principal customers were Cincinnati Milacron, Advanced Robotics Corporation, Cybotech, ASEA and Saipem (Italy). Merrick sales of its ThruArc Sensing System totaled only $3 million from 1981 to 1983. With only 10 percent of all arc welding done in the US, the system was quickly copied throughout the world, including Russia, China, Korea and Europe starting in 1984. Merrick did not have the resources to defend its four patents and sold the company to CRC Welding of Houston, TX that year. Total world-wide sales since 1984 of arc welding sensing systems (not including the robot) were estimated by the presidents of two leading arc welding sensing firms at about $700 million. They both estimated that at least 75 percent of all of these systems since 1984 use the through the arc Merrick method, or an estimated $525 million of indirectly related sales. Merrick research collaborations were with Vanderbilt, Cincinnati Milacron, Cybotec and ASEA. Cook said that "SBIR allowed Merrick to regain the leadership in the arc welding field through the NSF sensing technology project."

30. Spectron Development Laboratories, Inc., Formerly Costa Mesa, CA

Spectron received an NSF SBIR award in 1980 for research on an idea for Near Surface Flaw Detection by Ultrasonic Angle Imaging. The result was a breakthrough in airborne optical sensors and instruments, for which it became recognized world-wide in military and commercial aviation. The sensors of air data improved the capabilities and performance of aircraft and of safety in flight, including early recognition of air turbulence and windsheer. Other applications of sensors and instruments are in the electric power and gas industries. The former president stated that total sales directly related to NSF SBIR projects totaled $123 million. Major customers included Rockwell, General Dynamics, Ford, EPRI and Virginia Electric Power. The company received numerous other SBIR awards from other agencies and by the mid-1980's, had become the leading recipient of SBIR awards. Spectron was acquired by Titan Corporation in 1986. Employment went to a high of 100 in 1989 but thereafter declined steadily to insignificant today. Titan had redirected Spectron's R&D efforts toward proposals to secure large A4 Air Force contracts, which were not successful. This action irreparably damaged the Spectron's previous R&D and market leadership in the high performance aircraft sensors and instrumentation field. The company obtained 6 NSF related patents and had research collaborations with UC Berkeley, UC Irvine, Lockheed, EPRI and the Gas Research Institute. It also had 4 spinoffs, all of which received SBIR awards. These included Sigma, Metrolaser, Aerometrics and Statonics. The Spectron president stated that "SBIR was vital to providing the R&D resources for high risk/high payoff ideas and the ability to assemble a team of capable people." Also, he felt "The company probably would not have existed without SBIR."

31. Cyclotomics, Inc., Berkeley, CA

In 1981 NSF funded a research project entitled Coding for Band-Limited Channels and in 1982 Interleaved Coding for Bursty Channels. Both were aimed at the coming problems in satellite and digital communications and storage systems. The research resulted in important new decoding and modulation techniques that significantly improved the reliability of satellite communications and information storage while using substantially less bandwidth and higher density magnetic recording. This had a big impact on the reliability of satellite communications and the magnetic storage tape business. It also influenced the decoding of compact discs by a technique that is still in use today. This was of great interest to Kodak, which proceeded to acquire Cyclotomics in 1985. Cyclotomics sales totaled $5 million and licensee sales added another $9 million. However, the results led directly to Cyclotomics development of the optical disc CD ROM controller for Kodak. This was a significant technical achievement and resulted in sales of more than $100 million for Kodak. It was the biggest commercial success the Mass Memory Division of Kodak had ever had at the time. Major customers of Cyclotomics were Kodak, ITT, Spin Physics, Schugart, Xerox, Memorex and Burroughs. Employment increased from 9 to 40. Cyclotomics received 18 US and 9 foreign patents. Collaborations were with UC Berkeley, University of Southern California, Caltech, Ampex, Kodak, ONR, NRL and Lawrence Livermore. The president, a former full professor of electrical engineering at UC Berkeley, said that "The funding came at a critical time for the company and the developments would not have occurred without it. NSF and the SBIR awards also played a big role in the sale of the company discussions with Kodak." Kodak bought the company for $11.5 million and, 10 years after acquiring Cyclotomics, moved it to Rochester, NY in December 1995.

32. Aerodyne Research, Inc., Billerica, MA

Aerodyne is a leading firm in air pollution research, instruments, monitoring and control. It was the first to develop a commercial instrument that could reliably measure the green-house gas fluxes in the atmosphere and its effect on global warming. The first laser system to measure the on-road exhaust emissions by individual vehicles and aircraft engine emissions on the ground came from NSF SBIR research. So did a methane monitor which is used world-wide to monitor methane and identify sources of pollution. NSF related sales total $8 million with 10 percent exported. Customers have included GE, GM, Allied Signal, Westinghouse, Boeing, Honeywell, Hughes, TI, Rolls Royce, Boeing, McDonnell Douglas, GD, Grumman, Aerojet, Calspan, Lockheed, EPRI, Gas Research Institute, Chemical Manufacturers Association and 7 federal agencies. Aerodyne has licensed both hardware and software and spun off one company. Employment has dropped from 100 to 52 but 50 employees transferred with the spinoff. There have been 12 NSF related patents. Research collaborations have been with Harvard, Boston College, MIT, Utah State, New Hampshire, Princeton, Washington State, Battelle, Mitre, Rand, GRI, EPRI, CMA, IDA, JPL, NCAR, Sandia, and Lincoln, Lawrence Livermore and 5 NASA laboratories. Aerodyne's president said NSF's SBIR requirements changed the company from an exclusively R&D firm to one that also manufactures and provides services world-wide. It spurred the company to manufacture its 3 major instruments and to focus on atmospheric pollution, environmental and materials areas. Aerodyne refocussed its goals and changed from its early 90 percent Defense R&D contracting to 40 percent. Currently 30 percent of its business now commercial and 30 percent with non-Defense Federal agencies.

33. Verax Corporation, now Creative Biomolecules, Inc., Formerly in Lebanon, NH

Verax developed the first, long-life catalysts for immobilized microorganism fermentors, a breakthrough which allowed the continuous production of monoclonal antibodies and other high value medical proteins, such as Interferon, Interluken II and Tissue Plasminogen Activator from a 1982 NSF grant. Verax became a significant early producer and licensor of critically needed process bioreactor equipment which contributed to sharp reductions in the cost of these prohibitively expensive, but critically needed, medical and animal protein drugs. In a related NSF project, Verax developed innovative technology for the large-scale production of mammalian-cell proteins. It became a world leader in producing of mammalian cells for parentral biopharmaceuticals used by biotech firms and the pharmaceutical industry. The breakthroughs attracted $2 million from Eli Lilly which in turn attracted another $41 million from venture capital and large industrial firms. Verax sales totaled $25 million with 10 percent exported. Principal customers were Lilly, Merck, Green Cross (Japan) and Biogen. Employment grew from 12 to 125 before declining somewhat prior to its acquisition by Creative Biochemicals, Inc in 1993. Verax had obtained about 16 patents and had research collaborations with Dartmouth, Rutgers, MIT, Lilly, Sandoz, Biogen, Celltrix, Genzyme, Creative Biomolecules and Hemosol (Canada). The founder of Verax and principal investigator for the NSF projects left Verax in 1987 to co-found Synosis Corporation and serve as its chairman. He said that Synosis is a direct spinoff from Verax. It would not have happened without Verax, nor would Verax have succeeded without SBIR. Synosis also aimed at reducing the high cost of bioseparation and purification products but focused on improving chromatography instrumentation used to purify and analyze biomolecules. It grew rapidly, changed its name to PerSeptive Biosystems, Inc. and attracted $132 million from venture capital and public offerings. VC firms included Zero Stage Capital, 3i (Bank of England), Raytheon Ventures, Venrock, Bessemer, Morgan Holland, Copley, Highland and Prince. Total sales to date are $170 million.

PerSeptive Biosystems has 569 employees, more than 100 patents and research collaborations with Lilly, Boehringer-Mannheim, Myco and many small companies. The founder of both Verax and Synosis said that NSF and NIH both turned down his SBIR proposals from Synosis on the extremely successful concept because "the principal investigator was a mechanical engineer who obviously cannot improve chromatography." The founder of Verax said that "SBIR projects put Verax into the bio field, and therefore, PerSeptive Biosystems as well. He said "SBIR was the first outside money he had been able to obtain. It led directly to $2 million from Lilly which, in turn, led to $41 million more."

34. Optivision, Inc., Davis, CA

NSF funded an SBIR project on Optivision's basic concept which involved New Architectures and Optical Interconnections for Vision Systems. From it Optivision invented one of the first optical switching techniques that is still important in the field. It allowed them, with only about 100 employees founded in 1983, to beat out Rockwell, Unisys and three other firms for a Defense applications contract for all optical switching which enables the use of a high bandwidth. More recently they were also successful with SPRINT and MCI. Improved compression from another NSF award in 1991 has allowed low-cost transmission and high resolution still-frame and video to be transmitted over the telephone. A 1993 grant from NSF on Control Methods for Wide-Area All-Optical Networks is a recent significant breakthrough for Optivision in its fast changing field. "SBIR allowed us to try high risk ideas and our founders used the SBIR proposal and funding to start the company as planned. The exciting research also allowed us to attract top people. We would not be here without SBIR." The company received its first NSF SBIR award in 1984, the year following its founding. It had 3 employees at this time and now has 105. Major SBIR funding has also come from Defense and NASA. Sales total $14 million from five SBIR related product-lines with 40 percent exported, primarily to the UK, Germany, Italy and Japan. Principal customers are Kodak, IBM, Lockheed, Olicon, Silicon Graphics, Sun Microsystems, Microsoft, Phillips and Hewlett Packard. Optivision attributes the $6 million of industrial funding it has received to SBIR. It has received four patents and has had 10 research collaborations. These include Stanford, UC Davis, USC and two with Los Alamos. Optivision also is involved in a TRP with Hughes and others.

35. Brewer Science, Inc., Rolla, MO

Brewer was the first to develop chemistry for thin anti-reflective coatings for microfabricating electronic circuits to enhance high speed processing and increasing memory density. It also believes that it is the only company in the world, except for Japan firms, that has developed color filters and coatings technology for the flat panel industry. It was also the first to use dye-color polymer film for color filters that withstand high temperature processing. All come from 3 separate NSF SBIR funded research projects as with additional support from DOD and NIH. Brewer says that it has increased US competitiveness in flat panel displays and submicron electronic circuits with antireflectant coatings. Total sales exceed $30 million with 34 percent exported to Japan, Canada, Europe, Southeast Asia, Israel, Korea, Taiwan and South Africa. Customers include IBM, Motorola, National Semiconductor, NEC, Fujitsu, Matsushita, TI, HP, Sony, DEC, Siemens, Phillips, Plessey and GEC. The company has 120 patents and 34 collaborations which they attribute mostly to SBIR. These have been with the University of Wisconsin, SW Missouri State, Missouri-Rolla, Kent State, MIT, SEMATECH, Sandia, NRL, Rohm & Haas and 10 proprietary customers. Brewer "is proud that it is a significant producer of quality jobs in a semi-rural environment with SBIR's help. It has helped us create and retain high quality technical people. NSF provided the most fundamental research opportunities and the US Patent Office told us that one of our early patents was a 'hallmark' patent. It allows us to micromachine metal parts by removing the organic sacrificial layer without damaging the metal." The company has grown from 12 to 130 employees since its first NSF SBIR award in 1984.

36. Millitech Corporation, South Deerfield, MA

Millitech carried out NSF SBIR research and developed two instruments that were the first to be able to measure ozone and chlorine monoxide between 20-70 kilometers from a ground-based radio telescope. The instruments and technology are important contributions to long-term environmental monitoring of the world for long term ozone depletion. The company is a spinoff from the Five College Radio Astronomy Laboratory, a laboratory supported by NSF basic research grants. Millitech has produced 5 instruments at approximately $500,000 each, which came from NSF and NASA SBIR research projects. Sales total $2.5 million with 20 percent exported. Sales have been principally to the International Chemical Manufacturers Association and NASA. SBIR helped Millitech become the world leader in this small, very specialized, but important, ozone monitoring of the upper atmosphere instrumentation field. At the time the instruments contract was the largest program in the company. Employment has increased from 27 at the time of the NSF award to 130 today. Research collaborations have been with the University of Massachusetts and NASA Langley.

37. Sawyer Research Products, Inc., Eastlake, OH

Sawyer is a major US producer of high purity, electronic grade, quartz crystals. It is another example of how a single SBIR award made a major difference to a small high tech company. The resulting single crystal of cultured quartz represents one-half of the company's total sales and another quarter of their sales have benefitted significantly from the same research. SBIR related sales now total $69 million, 80 percent of which are exported. In the early 1980's Sawyer found that it had real problems because the quality of its crystals was becoming less competitive due to technological advances. The NSF Electrical, Computer and Systems Engineering Program was interested in the problem and funded their proposed research on The Role of Solid Phase Impurities in Dislocations and the Character of these Defects in Single Crystal Quartz. Phase I was promising and Phase II was supported for research on New Processes for the Growth of High Purity Electronic Grade Quartz Crystals. The research which the company could not otherwise afford resulted in an extremely high quality and competitive crystal for electronic needs. Proprietary customers include major electronics manufacturers both in the US and abroad and one of the major NSF SBIR successful firms. Most of the exports are to Germany, France, the UK and Korea. Collaboration was with Penn State, Case Western, Arizona State and Central Florida and with the Army Research Laboratory. Employment which had been 200 in 1985 is now 220. The president stated "SBIR made a huge difference to Sawyer and we had very significant results. The research helped us understand how to produce a defect-free pure single crystal."

38. Sievers Research Corporation, Boulder, CO

Sievers designed and developed a new type of gas chromatography detector that uses gold to catalyze the oxidation of hundreds of organic and inorganic compounds by nitrogen dioxide and synthesized oxidation products coupled with chemoluminescence measurements. The instrument has been very successful when used in chemistry, petrochemical research and in industrial process control. It also has major applications in the environmental field and has led to the development of other instruments. One, an NO2 detector measures sulfur in petrochemicals, medical blood flow and breath analysis. Another is a new water purifier that has great promise in the water quality-testing field. The water purifier also will go into space on MIR this December. Cumulative sales are $28 million with 40 percent exported to 35 countries. Customers include Intel, Motorola, Texas Instruments, Exxon, Amoco, Chevron and 20 more of the Fortune 100. Employment increased from 6 at the time of the first SBIR award to 100 at the time of sale of the company to Ionics, Inc. of Watertown, Massachusetts in May, 1996. There are 24 patents and three collaborations, the latter with the University of Colorado, Hewlett and with NASA for MIR. The NSF SBIR award was the first outside funding that Sievers' obtained and SBIR awards from NSF, DOE, NASA and EPA have financed their R&D ideas. Their SBIR success also helped them raise venture capital. With Sievers' acquisition by Ionics, for approximately $20 million, it looks forward to significant future growth, particularly in the water quality field. The new water quality measurement instrument will go into space on the Space Shuttle to Mir this December. The president, who is a Russian immigrant, and the founder, a University of Colorado chemistry professor, intended Sievers to be an R&D firm when was established in 1983. It was SBIR that focused them on a product and then becoming a commercial company.

39. AstroPower, Inc., Newark, DE

NSF and DOD SBIR research has led to superior thin layer, silicon and gallium-arsenide technology with optical and speed advantages in photovoltaic devices. It has made AstroPower the third largest US manufacturer and seventh largest in the world of photovoltaic devices with product sales totaling $19 million, 80 percent of which have been exported. NSF research projects were for Nucleation and Lateral Overgrowth of GaAs on Silicon for Optical Communications in 1986, for Integrated GaAs Emitter Element for Optical Interconnections in 1987, and for New Materials for Microelectronic and Optical Integrated Circuits in 1988. AstroPower products include the photovoltaic detectors, high performance UV and IR detectors, solar cells, IR energy conversion devices, light emitting diodes and specialty products for the Defense. Clients are original equipment manufacturers (OEM) world-wide, Niagara Mohawk Power and General Public Utility (NJ). Foreign customers have been in Germany, India, Spain, Australia and Mexico. SBIR has resulted in the company attracting additional private funding, venture capital and industrial investment totaling $6 million. At the time of the earl NSF awards AstroPower was a 30 employee division of AstroSystems, Inc. Today it has 130 employees. NSF research has resulted in or contributed to 31 patents. Collaborations have been with the University of Delaware, Georgia Tech, Sandia, DOE's National Renewable Materials Lab, Dow Chemical and Niagara Mohawk. The president said that "NSF and DOD support is responsible for the company's existence. This technology became the basis for our growth and our moves into other areas."

40. Key Curriculum Press, Inc., Berkeley, CA

NSF also supports research in science and engineering education. Key Curriculum has introduced a revolutionary approach to teaching geometry into 10,000 schools in 8 languages. It has conducted seven to ten day courses for approximately 8,000 teachers who have taught "Geometry Sketchpad"; these teachers reach over one million students in grade schools alone. The innovative technique involves teaching geometry on the computer with the computers unique capabilities to show relationships. Now 2,000 teachers a year are taking the courses. A second program is teaching probability and statistics in a visionary manner via "Dataspace" at the high school and college level in the US and world-wide. Dataspace also teaches students, teachers and the public how to access and use available, relevant government and other information on-line. A third program has replaced laboratory equipment in teaching optometry and Key Curriculum is exploring it as a model for teaching other science courses in the future. The company is extremely talented and very successful. The sales mostly relate to a 1990 NSF SBIR project completed in 1993 and they are already at $5 million. Sales are made through distributors in the US, Europe, Japan, Russia, Australia, China, Egypt and Saudi Arabia. Employment has increased from 12 to 60 during this period. Research collaborations have been with UC Berkeley, Carnegie-Mellon, Wright State, IBM, TI and a large German publisher. Key Curriculum says it attributes its success to NSF's willingness to fund advanced, cutting-edge, and innovative, teaching ideas. NSF's reputation and prestige has also helped immensely in creating alliances and partnering with major players in the education field.

41. Martek Corporation, Columbia, MD

Martek is a biotech firm principally involved in algae. A research breakthrough discovered has made it possible for a critical ingredient needed by babies, normally found in mothers milk, to be produced from algae. Martek says that nutritional organizations in Europe, including the World Health Organization, have recommended that the ingredient, referred to chemically as DHA, be added to infant formulas to offset nutrient deficiency in preterm babies and babies that are not breast fed. Currently Martek licenses 6 companies to use of DHA. They represent 40 percent of the world's infant formula sales. NSF SBIR was the first to provide Martek with research funds for a project on Biological Production of Novel Deuterated Lubricants in 1986. The results have been of fundamental importance to all Martek products. The NIH SBIR program has also made significant contributions to Martek's technical success. Martek's product sales to date are $10 million but this does not include the licensed sales which are just getting started in The Netherlands and Belgium. They will expand into most of Europe in 1996 and to the US in 1997. US clients include Mead Johnson, Wyeth and Nutritia. Other products include drug design reagents and aquiculture feed. The product appears to be very promising and Martek has been able to raise $77 million from venture capital and public offerings. The company has 117 patents and collaborations with 15 institutions which include Auburn, Arizona State, Washington, Maryland, Columbia, VPI, the Carnegie Institution and with one industrial firm, ISE. Martek, formerly Martin-Marietta's BioScience Department, spun off in 1985 and received its key NSF SBIR success award in 1986. According to Martek officials, "SBIR was the company's major support for R&D up until 1993. The NSF SBIR allowed us to convert an idea into technical results which were the key to raising $8 million in venture capital, $53 million from public offerings and $16 million from other sources. NSF was our earliest research support and it was fundamental to Martek's broad base in deuterated lubricants from algae and its raising of this investment."

42. Ecogen, Inc., Langhorne, PA

Ecogen accomplished a significant breakthrough in the environmental field with its discovery of the first rapid way to identify, clone and transfer genes for the development of a new bioinsecticide. Ecogen was successful and this has led to tripling the use of environmentally safe insecticide products. The breakthrough represents important progress and a promising possible answer to one of our major environmental problems, that of overuse of highly toxic, chemical insecticides. The breakthrough came from an NSF SBIR funded Ecogen proposal in 1986 on a Novel Cloning and Transfer Systems for Bacillus thuringiensis (BT) Insecticidal Genes. The aim was to identify the toxin gene, find a host bacteria, clone them and develop novel biological insecticides and maximum yields for potential commercial use. They did, discovered the innovative cloning methodology approach, and the results have had a significant impact on the bioinsecticide field. Sales of the NSF related product total $40 million, plus another $20 million of related non-government R&D. The results also have generated two, small joint ventures but they are too new for sales. One is with Monsanto and the other with Rousell Ucles (France). The project also helped Ecogen to obtain $39 million of outside investment including $9 million in venture capital and $27 million from a public offering. Employment has increased from 41 at the time of the first award to about 100 today. Ecogen also has 27 related patents and one research collaboration with Seregen (Monsanto). The company vice-president for R&D said, "The NSF project was very important. Ecogen wouldn't be here today without it. SBIR research results are the basis of the company's business".

43. CyberOptics, Inc., Golden Valley, MN

NSF funded an SBIR project with CyberOptics in 1987 on Non-Contact Pencil Probe for Advanced Manufacturing Inspection to help meet the need for more rapid and accurate inspection techniques. The research aimed at the ultimate development of a high accuracy, miniature pencil probe capable of inspecting manufactured parts with complex geometries. Although the project was not entirely successful commercially, the research was ahead of others in the market, on target and fundamentally important to the company's rapid growth in the sensor field. CyberOptics is now the innovation and technology leader in this electronics niche. Cumulative sales are $70 million with a 100 percent growth last year. Some 67 percent of sales are exported about equally to Europe and Japan. Its laser-based, high precision sensors products are mostly sold OEM to major electronics firms, such as Phillips, as well as to NASA. It also has a CyberScan Station which is to the same market. The company had a small initial public offering in 1988 and a $37 million one last year. Employment has increased from 12 at the time of the NSF award to 200 today. The company has 57 patents and has had 12 research collaborations with the University of Minnesota where the president was formerly a professor, IBM, Motorola, TI, Phillips, Siemens, Fuji, Yamaha, Panasonic, Samsung, Aman and a German research institute. The CyberOptics president said that "The NSF award was important in 1987 because the company had little money from its founding in 1984 and was faced with the stock market collapse at that time. SBIR was right on target for CyberOptic's focus and the NSF credibility helped with the stock offering in March of 1988."

44. Displaytech, Inc., Boulder, CO

Displaytech is a leading firm in developing color miniature liquid crystal displays based upon ferroelectric liquid crystal (FLC) technology that has significant potential in many emerging communications areas. The company has produced realistic full color pictures at video-rate speeds that are high resolution, lightweight and low in power consumption as well as ChronoColor miniature displays for optical document readers, personal viewers, front and rear projections and optical correlations for fingerprint reading. Sales of components exceed $2 million and the NSF results were a key to the company receiving an Advanced Technology Program (ATP) award. NSF SBIR projects have been important to Displaytech's achievements. NSF made its first and basic grant for Synthesis and Evaluation of New Ferroelectric Liquid Crystals in 1987 plus other research projects relating to high speed applications, ultrahigh polarization of FLC's, FLC's for infrared modulators and for the linear electroptic effect of sinectic analog materials. Displaytech has attracted $3 million in venture capital and employment has increased from 6 at the time of their first SBIR award to 42. The company has 10 related patents and research collaborations with the University of Colorado and MIT. It has also had SBIR funding from Defense and NASA.

45. Communication Intelligence Corporation, Redwood Shores, CA

CIC is already a world leader in the revolutionary development of pen computing products, such as handwriting recognition for signature verification, computer entry products, the editing of computer text and related software, in part as a result of NSF SBIR research. This includes CIC leadership in understanding human behavior characteristics in using the pen to instruct the computer. NSF SBIR projects with CIC since 1988 include: (1) On-Line Cursive Handwriting Recognition, (2) OCR Handwriting Recognition with Application to Man-Machine Interaction, and (3) Personal Authentication for Database and Network Security. Pen computing is a rapidly emerging new technology that may have many potential applications. For example, handwritten editing of computer generated text and graphics and data security are rapidly growing new markets. CIC says that it wants to bring multi-lingual "pen computing" to the world. NSF's SBIR program has contributed to CIC's surprising start. Fifty percent of all major computer manufacturers in the world, and 80 percent of Japan's, already use CIC handwriting recognition products and software. Major customers include IBM, Apple, NCR, Lotus, Epson, Fujitsu, Samsung and ACER. Chase Manhattan and GNMA are recent converts. The president of CIC said that "The SBIR awards came at a critical time for CIC in terms of the market opening and the company's need for financing after going through its own $1 million. We had not been able to attract other funding for this very high risk idea. The projects were of major importance to their technical success and the results are used on all of the company's products to date. NSF credibility also helped CIC raise additional money and obtain partners and collaborations. CIC now licenses its software to 30 computer manufacturers." Total CIC licensee sales of the software and hardware directly related to the pen/computer interface (only) are more than $1.5 billion world-wide. CIC's directly related private investment is $51 million to date. Employment has increased from 12 to 51. (CIC operates like a virtual corporation and farms out all but the critical R&D.) The company has 37 patents and nine collaborations. The latter include UC Berkeley, Stanford, Waterloo (Canada), NCR, Apple, IBM, Samsung and ACER. CIC also has been the recipient of a NIST ATP project in this field.

46. Aerometrics, Inc., Sunnyvale, CA

Aerometrics is the developer and manufacturer of the world's leading Phase Doppler Analyzer. SBIR awards from NASA, NSF and DOD have contributed to the company's strength in the marketplace. SBIR has allowed Aerometrics to pursue major advances to increase its performance which otherwise would not have been possible because of the cost and risk involved. NSF funded projects for diagnosing swirl combustion processes, diagnosing the effects of turbulence structure on cavitation of turbomachinery, and using laser diagnostics for interacting droplet combustion in low-grade fuels. Our results have made a major contribution to each of these areas. Aerometrics products are used world-wide for weather, pollution formation in the environment, combustion measurement, engine efficiency, etc. For example, gas turbine efficiency has improved performance and pollution reduction. 90 percent of automotive companies use their equipment world-wide for fuel injection measurement and 80 percent of national energy production is via fuel spray. According to company officials, "75 percent of the company's current sales are due to improvements that would not have been possible without SBIR. SBIR funds ideas. Sales total $45 million, half are due to SBIR but only $7 million directly to NSF." Major customers are the automotive, aerospace and fuel spray nozzle manufacturers, the power and energy industry, and universities and national laboratories for research. About 55 percent of sales are from overseas, principally Japan, Germany, Korea, France, Italy and the UK.

47. Altus Biologics, Inc., Cambridge, MA

NSF made an award in 1992 to Vertex Pharmaceuticals, Inc. for Crosslinked Enzyme Crystals as Catalysts in Organic Synthesis. The objective was to determine if crosslinked enzyme catalysts (CLEC) is a technology for stabilizing enzymes so that they can work in extreme environments used in industrial chemical processing. An Eli Lilly team said the CLEC developed under the NSF SBIR project was one of the most significant breakthroughs in biocatalysts of the last 10 years. Vertex spun the project off in 1992 and formed Altus Biologics. By 1995 Altus had 3 products which were sold to over 100 companies in 15 countries. In 1996 it has 7 products sold to 180 companies in 25 countries. Sales total $3 million with 60 percent exported, mostly to the UK, France and Germany. Major customers include Lilly, Hofmann LaRoche, Bayer, Ciba Geigy, ChiroScience, Eastman Chemical and Vertex itself, which used it to come up with a new anti-cancer drug. The project has generated $10 million in capital from Vertex. Altus has 5 patents and had 6 collaborations to date. These include Texas A&M, Pittsburgh, Glaxo-Welcome, Ciba-Geigy, 3M and the Army's Engineering R&D Center at Aberdeen, MD. The president of Altus said, "NSF had faith in our concepts and provided funds for early stage development of the technology when Altus only had 3 employees and no products or revenues. As a result of NSF's support, we were able to obtain funding and expand our efforts. We now have 15 employees and expect to approach $1 million in revenues this year."

48. Computer Motion, Inc., formerly Dynamic Microsystems, Goleta, CA

Computer Motion states that, as of 1996, it had the only surgical robot in the world cleared by FDA for operating room use. The company was the first to develop a 6 axis robotic arm, and the first to combine the robot, computer and camera into a laparascopic camera for minimally invasive surgery. It is used principally for surgery on the gall bladder, appendectomies, tube tying, and endometriosis. Benefits of the approach include less risk, less pain, smaller scar, quicker recovery, less cost, less time in the hospital, and lower time burdens on doctors and staff. The actual quality of surgery is considered to be better because the doctor can see better with the laparascope and the picture on the monitor. Computer Motion was formed in 1988 by the first graduate of the NSF Robot Center at UC Santa Barbara and it was the first successful spinoff from the Robotics Center and received an NSF SBIR award the next year. Sales of about 200 robots already exceed $4 million with 10 percent exported to Europe and Japan. The company has raised $11 million of venture capital, has 6 patents and 6 research collaborations. The latter are with Johns Hopkins, George Washington, Yale, Emory, St. Joseph and the Cleveland Clinic. The company now has 65 employees compared to 5 at the time of proposal. Computer Motion obtained 10 of the employees from the Robotics Center when it was terminated. The company's president says that "The NSF funding was critical early financing and it served as preventure capital. Without SBIR, it is highly unlikely that the company would exist." SBIR funding has also been received from NIH and NASA.

49. NonVolatile Electronics, Inc., Eden Prairie, MN

NonVolatile is the first company to make giant magneto resistance sensors, which are important to automotive markets for speed, position and current detection relative to brakes, anti-lock brake systems, engine management (cam and crankshaft position), position sensing of machine tools. In industrial markets, the sensors are used for automatic utility meter readers. The firm received an NSF SBIR research award in its first year of operation on Highly Magnetoresistive Thin Film Multilayers and another from NASA proved to be basic to the company's new technology and its rapid acceptance in the marketplace. Sales total $5 million and customers include ITT (brakes), Phillips (engine management), Wabash (machine tool positioning) and Itron (automatic utility readers). This generated $3 million in venture capital from Norwest Venture Capital and Motorola. NonVolatile also has initiated 4 licenses, which are just starting, and the company has is a recently hired distributor in the UK. Employment has gone from one plus 3 part-time at the time of NSF's 1990 award to 53 at this time. The company has 16 patents and 12 research collaborations. The latter include the University of Minnesota, Iowa State, MIT, CMU, Alabama, Stanford, NIST, Lawrence Livermore, Southwest Research Institute, Honeywell, Seagate and Motorola.

50. Applied Science & Technology, Inc., Woburn, MA

The company, which also is known as ASTeX, has had a major success from two NSF SBIR awards, one on Microwave Plasma Diamond Deposition at High Growth Rates and the other on Plasma and Ultraviolet Damage in Oxygen Photoresist Stripping with a Downstream Microwave Plasma Source. The resulting product from two Phase I awards was so successful that sales exceeded $4 million after only two years on the market and should continue to grow. SBIR and ARPA provided much of ASTeX's core technology used in its production of diamond films using the chemical deposition process. Directly and indirectly $10 million of sales were attributed to the base technology from NSF and ARPA SBIR research. About 60 percent of sales are exported to Japan, Hong Kong, Korea, Taiwan, Germany, the UK and Russia. Major products are plasma deposition and diamond deposition systems. Major customers include Mitsubishi, Kobe Steel, GE and GEC (UK). The company says that SBIR success has also contributed to its obtaining $21 million of investment, $16 million from a public offering, $4.5 million from venture capital, and $500,000 from private investment. Employment was 52 in 1991 at the time of the first award. It is 260 today, including 90 that came with the acquisition of another firm. ASTeX has 12 related patents and 7 collaborations. The collaborations include Boston University, NC State, Penn State, MIT, Oak Ridge, Sandia and Applied Materials.

Annex C

Participants List*

Zoltan Acs
University of Maryland

Pablo Amor
Delegation of the European Commission

Todd Anuskiewicz
MERRA

Martin A. Apple
Council of Scientific Society Presidents

Gene Banucci
Advanced Technology Materials, Inc.

Jon Baron
U.S. Department of Defense

Brian Belanger
NIST

Terry Bibbens
U.S. Small Business Administration

Joseph Bishop
U.S. Department of Commerce

Grant Black
Georgia State University

Brenda Black
U.S. Department of Commerce

Jeff Bond
U.S. Department of Defense

Jeffrey Brancato
National Science Foundation

Kelly Brickell
U.S. Department of Commerce

Peter Cahill
BRTRC

*Italics indicate presenters and panelists.

Robert K. Carr
SRI International

Richard W. Carroll
Digital System Resources, Inc.

Dennis Chamot
National Academy of Sciences

Charles Cleland
U.S. Department of Agriculture

William Colglazier
National Research Council

Paul Cooksey
Senate Committee on Small Business

Toni L. Cooper
Office of Representative John F. Tierney

Ritchie Coryell
National Science Foundation

Laureen Daly
Office of Senator Joseph Lieberman

Lance A. Davis
U.S. Department of Defense

The Honorable Tom Davis
Member of Congress

Arlene M. de Blanc
U.S. Department of Energy

Tony Diamond
National Aeronautics & Space Administration

Marshall Dick
U.S. Environmental Protection Agency

Jill Dickman
U.S. Air Force

Bruce Don
RAND Corporation

Ann Donowell
Office of Senator John Kerry

John Dougherty
The Illinois Coalition

E. Martin Duggan
Small Business Exporters Association

Paul Eckert
Office of Senator John Breaux

Lee Eiden
U.S. Department of Education

Art Engels
The Ram Company

Ann Eskesen
Innovative Development Institute

The Honorable Martyn Evans
Australian Parliament

Kenneth Flamm
The Brookings Institution

Pamela Flattau
Flattau Associates, LLC

Dick Fleming
The Ram Company

Patty Forbes
Senate Committee on Small Business

Kenneth Gabriel
U.S. Air Force

James Gallup
U.S. Environmental Protection Agency

Marc Garufi
Office of Management and Budget

Scott E. Geesey
House Committee on Science

Pam Guzzone
National Aeronautics & Space Administration

Victoria Hadfield
SEMI

Bruce K. Hamilton
National Science Foundation

James Hansley
USAID

Bo Hedberg
Embassy of Sweden

Joseph Henebury
U.S. Department of Transportation

Paul Herer
National Science Foundation

Michael Hettinger
Office of Congressman Tom Davis

Daniel O. Hill
U.S. Small Business Administration

Brian C. Holloway
Office of Senator John D. Rockefeller

Sara Horrigan
Office of Management and Budget

Dean Hudson
U.S. Army Research Lab

Connie Jacobs
Defense Advanced Research Projects Agency

John E. Jankowski
National Science Foundation

Maryellen Kelley
National Institute of Standards and Technology

Norman H. Kreisman
U.S. Department of Energy

Herbert J. Kreitman
National Institutes of Health

Witold Kuncewicz
DEMACO

Patrice Laget
Delegation of the European Commission

Clifford E. Lanham
Technology Planning and Management

Charles F. Larson
Industrial Research Institute, Inc.

Lena Lawrence
National Research Council

Josh Lerner
Harvard University

Lee Mercer
National Association of Small Business Investment Companies

Stephen A. Merrill
National Research Council

Duncan Moore
The White House
Office of Science & Technology Policy

Gary J. Morgenthaler
Morgenthaler Ventures

Mark Myers
Xerox Corporation

Kesh S. Narayanan
National Science Foundation

Robin Nazarro
U.S. Government Accounting Office

Robert L. Neal
U.S. Department of Defense

Allen Neece
Neece, Cator & Associates, Inc.

Carl Nelson
Carl Nelson Consulting, Inc.

Sara B. Nerlove
National Science Foundation

Robert L. Norwood
National Aeronautics & Space Administration

David O'Hara
Parallax Research, Inc.

John C. Oldfield
National Research Council

Karen L. Pera
U.S. Special Operations Command

John Preston
Quantum Energy Technologies

Teresa Puretz
U.S. Department of Defense

Andee Rappazzo
BDM International

Robert J. Reynik
National Science Foundation

James Roberts
American Agribusiness Equipment

Matthew Roberts
BRTRC

Dorothy Robyn
The White House

Daniel Roos
Massachusetts Institute of Technology

Charles Rowe
House Committee on Small Business

Richard Russell
House Committee on Science

Verl Sanders
U.S. Department of Health and Human Services

R. Wayne Sayer
Wayne Sayer & Associates

Marcia A. Smith
Columbia Cascade, Inc

David Speser
Foresight Science & Technology Inc.

Richard Spivack
National Institute of Standards and Technology

Norman Taylor
National Institute of Standards and Technology

Richard Thayer
TTI, Inc.

Roland Tibbetts
National Science Foundation

James H. Turner
House Committee on Science

Christine Villa
BRTRC

Andrew Wang
National Institute of Standards and Technology

Charles W. Wessner
National Research Council

John Williams
U.S. Department of the Navy

Jack Yadvish
National Aeronautics & Space Administration

Gary Young
American Xtal Technology

Jean Zettler
National Technology Transfer Center

Annex D

Bibliography

Acs, Z., and D. Audretsch, 1991. *Innovation and Small Firms*. MIT Press: Cambridge, Mass.

Audretsch, D., 1995. *Innovation and Industry Evolution*. MIT Press: Cambridge, Mass.

Audretsch, D., and R. Thurik, 1999. *Innovation, Industry, Evolution, and Employment*. Cambridge University Press.

Archibald, R. and D. Finifter, 1999. "Evaluation of the DoD SBIR Program and the Fast Track Initiative: A Balanced Approach," in *The Small Business Innovation Research Program: An Assessment of DoD's Fast Track Initiative*. Washington, D.C.: National Academy Press.

Barfield, C. and W. Schambra, eds., 1986. *The Politics of Industrial Policy.* Washington, D.C.: American Enterprise Institute for Public Policy Research.

Barry, C.B., 1994. "New directions in research on venture capital finance," *Financial Management* 23 (Autumn), 3-15.

Bingham, R., 1998. *Industrial Policy American Style: From Hamilton to HDTV*. New York: M.E. Sharpe.

Birch, D., 1981. "Who Creates Jobs," *The Public Interest,* Vol. 65 (Fall), 3-14.

Branscomb, L., and J. Keller, 1998. *Investing in Innovation: Creating a Research and Innovation Policy.* Cambridge, Mass.: MIT Press.

Brav, A., and P.A. Gompers, 1997. "Myth or reality?: Long-run underperformance of initial public offerings: Evidence from venture capital and nonventure capital-backed IPOs," *Journal of Finance*, 52, 1791-1821.

Brown, G., and J. Turner, 1999. "Reworking the Federal Role in Small Business Research," *Issues in Science and Technology,* vol. XV, no. 4, Summer.

Cahill, P., 1999. "Fast Track: Is It Speeding Commercialization of DoD SBIR?" in *The Small Business Innovation Research Program: An Assessment of DoD's Fast Track Initiative.* Washington, D.C.: National Academy Press.

Coburn, C., and D. Bergland, 1995. *Partnerships: A Compendium of State and Federal Cooperative Technology Programs*. Columbus, OH: Battelle.

Cohen, L.R., and R.G. Noll, 1991. *The Technology Pork Barrel.* Washington, D.C.: The Brookings Institution.

Council of Economic Advisers, 1995. *Supporting Research and Development to Promote Economic Growth: The Federal Government's Role.* Washington, D.C.

Davis, S.J., J. Haltiwanger, and S. Schuh, 1994. "Small Business and Job Creation: Dissecting the Myth and Reassessing the Facts," *Business Economics*. Vol. 29, no. 3, 113-22.

Flamm, K., 1988. *Creating the Computer*. Washington, D.C.: The Brookings Institution.

Freear, J., and W.E. Wetzel, Jr., 1990. Who bankrolls high-tech entrepreneurs?, *Journal of Business Venturing* 5, 77-89.

Galbraith, J.K., 1957. *The New Industrial State*. Boston: Houghton Mifflin.

Gompers, P.A., 1995. Optimal investment, monitoring, and the staging of venture capital, *Journal of Finance* 50, 1461-1489.

Gompers, P.A., and J. Lerner, 1996. "The use of covenants: An empirical analysis of venture partnership agreements," *Journal of Law and Economics* 39, 463-498.

Gompers, P.A., and J. Lerner, 1999. "An analysis of compensation in the U.S. venture capital partnership," *Journal of Financial Economics*, vol. 51, no. 1, pp. 3-7.

Gompers, P.A., and J. Lerner, 1998. What drives venture capital fund-raising? (Unpublished working paper, Harvard University).

Graham, O.L. 1992. *Losing Time: The Industrial Policy Debate*. Cambridge, Mass.: Harvard University Press.

Griliches, Z., 1990. *The Search for R&D Spillovers*. Cambridge, Mass.: Harvard University Press.

Hall, B.H., 1992. Investment and research and development: Does the source of financing matter? (Working Paper No. 92-194, Department of Economics, University of California at Berkeley).

Hill, C., 1998. "The Advanced Technology Program: opportunities for enhancement," in Lewis Branscomb and James Keller, eds. *Investing in Innovation: Creating a Research and Innovation Policy*. Cambridge, Mass.: MIT Press. 143-173.

Horrigan, J., 1999. Cooperating Competitors: A Comparison of MCC and SEMATECH. Monograph. Washington, D.C.: National Research Council.

Kleinman, D.L., 1995. *Politics on the Endless Frontier: Postwar Research Policy in the United States*. Durham, N.C.: Duke University Press.

Krugman, P., 1990. *Rethinking International Trade*. Cambridge, Mass.: MIT Press.

Krugman, P., 1991. *Geography and Trade*. Cambridge, Mass.: MIT Press.

Lebow, I., 1995. *Information Highways and Byways: From the Telegraph to the 21st Century*. New York: Institute of Electrical and Electronic Engineering.

Lerner, J., 1994. "The syndication of venture capital investments," *Financial Management* 23 (Autumn), 16-27.

Lerner, J., 1995. "Venture capital and the oversight of private firms," *Journal of Finance* 50, 301-318.

Lerner, J., 1998. "Angel financing and public policy: An overview," *Journal of Banking and Finance*, vol. 22, no. 6-8, 773-784.

Lerner, J., 1999. "The government as venture capitalist: The long-run impact of the SBIR program," *Journal of Business*. July, v.72, 3, pp. 285-97..

Link, A.N., 1998. "Public/Private Partnerships as a Tool to Support Industrial R&D: Experiences in the United States." Paper prepared for the working group on Innovation and Technology Policy of the OECD Committee for Science and Technology Policy, Paris.

Long, W.F., 1999. *Performance of completed projects: Status report number 1*. Washington, D.C.: U.S. Department of Commerce, NIST Special Publication 950.

Malone, T., 1995. *The Microprocessor: A Biography*. Hamburg, Germany: Springer Verlag/Telos.

Mansfield, E., 1985. "How Fast Does New Industrial Technology Leak Out?" *Journal of Industrial Economics*. Vol. 34, no. 2.

Mansfield, E., 1996. *Estimating Social and Private Returns from Innovations Based on the Advanced Technology Program: Problems and Opportunities*. Unpublished report.

McCraw, T., 1986. "Mercantilism and the market: antecedents of American industrial policy," in Barfield, C. and Schambra, W., eds., *The Politics of Industrial Policy*. Washington, D.C.: American Enterprise Institute for Public Policy Research.

Mowery, D., ed., 1999. *U.S. Industry in 2000: Studies in Competitive Performance.* Washington, D.C.: National Academy Press.

Mowery, D., 1998. "Collaborative R&D: how effective is it." *Issues in Science and Technology* Fall, 37-44.

Mowery, D., and N. Rosenberg, 1998. *Paths of Innovation: Technological Change in 20th Century America.* New York: Cambridge University Press.

Mowery, D., and N. Rosenberg, 1989. *Technology and the Pursuit of Economic Growth.* New York: Cambridge University Press.

Myers, S., R.L. Stern, and M.L. Rorke, 1983. A study of the Small Business Innovation Research program, Lake Forest, Illinois. Mohawk Research Corporation.

National Research Council, 1996. *Conflict and Cooperation in National Competition for High-Technology Industry.* Washington, D.C.: National Academy Press.

National Research Council, 1997. *Review of the Research Program of the Partnership for a New Generation of Vehicles: Third Report.* Washington, D.C.: National Academy Press.

National Research Council, 1998. *Effectiveness of the Advanced Battery Consortium as a Government-Industry Partnership,* Washington D.C.: National Academy Press.

National Research Council, 1999. *Funding A Revolution: Government Support for Computing Research.* Washington D.C.: National Academy Press.

Nelson, R.R., 1982. *Government and Technological Progress.* New York: Pergamon.

Nelson, R.R., 1986. Institutions supporting technical advances in industry. *American Economic Review, Papers and Proceedings,* 76(2):188.

Nelson, R.R., ed. 1993. *National Innovation System: A Comparative Study.* New York: Oxford University Press.

Organization for Economic Cooperation and Development, 1997. *Small Business Job Creation and Growth: Facts, Obstacles, and Best Practices.* Paris.

Perret, G., 1989. *A Country Made by War: From the Revolution to Vietnam—The Story of America's Rise to Power.* New York: Random House.

Price, Waterhouse, 1985. Survey of Small High-tech Businesses Shows Federal SBIR Awards Spurring Job Growth, Commercial Sales. Washington, Small Business High Technology Institute.

Romer, P. 1990. Endogenous technological change. *Journal of Political Economy* 98:71–102.

Rosenbloom, R., and W. Spencer, 1996. *Engines of Innovation: U.S. Industrial Research at the End of an Era.* Boston: Harvard Business Press.

Sahlman, W.A., 1990. "The structure and governance of venture capital organizations," *Journal of Financial Economics* 27, 473-521.

Schumpeter, J., 1950. *Capitalism, Socialism, and Democracy.* New York: Harper and Row.

Stowsky, J., 1996. "Politics and Policy: The Technology Reinvestment Program and the Dilemmas of Dual Use." Mimeo, University of California.

U.S. Congressional Budget Office, 1985. *Federal financial support for high-technology industries* (Washington, U.S. Congressional Budget Office).

U.S. General Accounting Office, 1987. *Federal research: Small Business Innovation Research participants give program high marks* (Washington, U.S. General Accounting Office).

U.S. General Accounting Office, 1989. *Federal research: Assessment of Small Business Innovation Research program* (Washington, U.S. General Accounting Office).

U.S. General Accounting Office, 1992. *Federal research: Small Business Innovation Research shows success but can be strengthened* (Washington, U.S. General Accounting Office).

U. S. General Accounting Office, 1999. *Federal Research: Evaluations of Small Business Innovation Research Can Be Strengthened,* (RCED-99-198).

U.S. Senate Committee on Small Business, 1981. Senate Report 97-194, *Small Business Research Act of 1981,* September 25, 1981.

U.S. Small Business Administration, 1994. *Small Business Innovation Development Act: Tenth-year results* (Washington, U.S. Government Printing Office).

Wessner, C., ed., 1999a. *A Review of the Sandia Science and Technology Park Initiative.* Washington, D.C.: National Academy Press.

Wessner, C., ed., 1999b. *The Advanced Technology Program: Challenges and Opportunities*. Washington, D.C.: National Academy Press.

Wessner, C., ed., 1999c. *New Vistas in Transatlantic Science and Technology Cooperation*. Washington, D.C.: National Academy Press.

Wessner, C., ed., 1999d. *The Small Business Innovation Research Program: An Assessment of DoD's Fast Track Initiative*. Washington, D.C.: National Academy Press.